THE MEDIUM PICTURE

Also by Roy Christopher

AS AUTHOR

The Grand Allusion: A New Understanding of Popular Culture

Post-Self: Journeys Beyond the Human Body

Different Waves, Different Depths: Stories

Abandoned Accounts: Poems

Dead Precedents: How Hip-Hop Defines the Future

AS EDITOR

Boogie Down Predictions: Hip-Hop, Time, and Afrofuturism

Follow for Now: Interviews with Friends and Heroes

Follow for Now, Vol. 2: More Interviews with Friends and Heroes

Advance Praise

"Exactly the sort of contemporary cultural analysis to yield
unnerving flashes of the future."
—William Gibson, author of *The Peripheral*

"Like a skateboarder repurposing the utilitarian textures of
the urban terrain for sport, Roy Christopher reclaims the content and
technologies of the media environment as a landscape to be navigated
and explored. *The Medium Picture* is a highly personal yet revelatory
chronicle of a decades-long encounter with mediated popular culture."
—Douglas Rushkoff, author of *Team Human*

"If you want to understand the social, psychological, cultural effects
of the media explosions of the past 50 years, *The Medium Picture*—
thoughtful, comprehensive, and deep—is for you. Immersed in the
contemporary digital culture he grew up with as a teenager, Roy
Christopher is old enough to recall vinyl, punk, and zines—social media
before TikTok and smartphones. *The Medium Picture* deftly illuminates
the connections between post-punk music critique, the increasing
virtualization of culture, the history of formal media theory, the liminal
zones of analog versus digital, pop versus high culture, capitalism versus
anarchy. It's the kind of book that makes you stop and think
and scribble in the margins."
—Howard Rheingold, author of *Net Smart*

"A synthesis of theory and thesis, research and personal recollection, *The
Medium Picture* is a work of rangy intelligence and wandering curiosity.
Thought-provoking and a pleasure to read."
—Charles Yu, author of *How to Live Safely in a Science Fictional Universe*

"What we need is a parkour for the Interzone triangulated by 'street' culture, the poets laureate of digital disruption, and the hairier fringes of academic discourse. If anybody's going to take a crack at it, Roy Christopher—certified skater boy, PhD in communication studies, fractal geometer tracing the coastlines of subcultural thought—is that thinker."
—Mark Dery, author of *The Pyrotechnic Insanitarium*

"If the medium is the message, then today's mediascape is a constant digital missive, reminding us that we are trapped together in a seven-level underworld metaverse mall. Who better to help guide us through this neo-Dantesque pixel inferno than Roy Christopher— one part Virgil, one part Debord, and at least one part his mischievous and brilliant self."
—Dominic Pettman, author of *Infinite Distraction*

"Through music, generational habits, pre- and post-internet cultures, and a multitude of 'cognitive entanglements,' this book flows with grace across different scales of mediation and affect."
—Jussi Parikka, author of *A Geology of Media*

"Brilliant, pathbreaking, palpable insights . . . Worthy of McLuhan."
—Paul Levinson, author of *New New Media*

"This book looks wonderful!"
—Laurie Anderson

THE MEDIUM
PICTURE

ROY CHRISTOPHER

THE UNIVERSITY OF GEORGIA PRESS
ATHENS

© 2025 by the University of Georgia Press
Athens, Georgia 30602
www.ugapress.org
All rights reserved
Designed by Melissa Buchanan
Set in Garamond Premier Pro and Trade Gothic

Most University of Georgia Press titles are
available from popular e-book vendors.

Printed digitally

Library of Congress Cataloging-in-Publication Data
Names: Christopher, Roy (Editor) author
Title: The medium picture / Roy Christopher.
Description: Athens : The University of Georgia Press, [2025] |
Includes bibliographical references and index.
Identifiers: LCCN 2025019709 | ISBN 9780820374345 hardback |
ISBN 9780820374352 paperback | ISBN 9780820374369 epub |
ISBN 9780820374376 pdf
Subjects: LCSH: Technology—Social aspects | Technological innovations—
Social aspects | Interpersonal communication
Classification: LCC T14.5 .C495 2025
LC record available at https://lccn.loc.gov/2025019709

In memory of Dave Allen

It's a far cry from the world we thought we'd inherit,
And it's a far cry from the way we thought we'd share it.

—Rush, "Far Cry"

How, from a fire
that never sinks
or sets,
would you escape?

—Heraclitus, "Fragment 27"

CONTENTS

ILLUSTRATIONS

ILLUSTRATIONS

FOREWORD

Exceeding My Grasp, or, In One Era, Out the Other

Andrew McLuhan

If only life were like this.

—Woody Allen as Alvy Singer
in *Annie Hall*

It matters that Marshall McLuhan was not Woody Allen's first choice for the memorable *Annie Hall* cameo (the pun is, if not always intended, always taken advantage of). It also matters that he ended up casting him and using him in the film. Relatable. Marshall McLuhan (my grandfather), or Eric McLuhan (my father), for that matter, were not my first choice either, but here I am, and here we are, wherever and whenever that is. I'm stretching the metaphor, which I suppose is the point of a metaphor in the first place ("a man's reach must exceed his grasp or what's a metaphor?"), but hopefully not past the breaking point. Maybe I'm just trying to say that life is weird and funny and that, try as we might to avoid them, certain things seem destined.

Roy Christopher and I, and I suspect you also, dear reader, have many things in common. At various times in my life, I've been a poet, a punk rocker, a skate-boarder. I've also had an enduring love for hip-hop and even a brief stint as a would-be performer ("talk about it, be about it"). I'm still fascinated by the art of beat juggling and from time to time mess around and try to learn the skill. It's much harder than it looks.

Another thing we have in common is a drive to explore and make sense by diverse means. There's little linear about it except maybe that we get into a groove from time to time but tend to skip and drop the needle with the abandon and joy of both explorer and navigator. Perhaps we have goals and destinations but are not so concerned with exactly how we reach them or even if we end up somewhere completely different than we imagined.

It could be that we're less concerned with making sense than with experience, but that's not to say that we're unconcerned with making sense. There's a tension there between experience and understanding, which I think many of us feel keenly. If you put the two things on a scale, it's interesting to consider how the two weigh relative to each other, and how that relation changes.

As I age, I find the scales tip toward making sense.

Then again, to separate sense and experience may be to miss their interrelation. After all, making sense is the end of an operation that throws all our experience into a blender or mixing board, resulting in something that just sounds, tastes, feels, right—that makes sense. Kind of like this book.

If you weren't trying to make sense, you wouldn't be reading this. In another sense, all media make sense, which has much (or everything) to do with both experience and comprehension. In a very real way, media make the means by which we travel and arrive anywhere. We are processed and reprocessed by and in the process.

In the following pages, the author brings together and builds a galaxy of influence and interest, like a turntablist working the crowd. Choice cuts and samples are deftly dropped to build and release tension. At the end of the show, I think we'll all go away a little exhausted but with smiles on our faces. We might pay for it in the morning, but we'll be better for it and without regret.

PREFACE

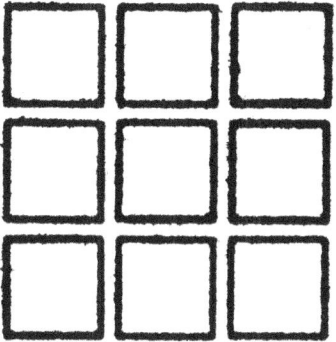

You see people, these days, who give the impression that their minds are a complete vacuum; no dreams or hopes of any importance—even to themselves—emanate through the sutures of their skulls, as it were. But that doesn't matter, in a sense, because the environment itself is doing the dreaming for them. The environment is the greater sensorium generating these individual hopes and ambitions, signs of the cerebral activity that has been transferred from inside the individual's skull into the larger mental space of the planetary communications landscape.

—J. G. Ballard

Why do we attend to the things to which we attend?

—James Ten Broeke

Western man is externalizing himself in the form of gadgets.

—Dr. Benway, in William Burroughs's *Naked Lunch*

A BAD MIRACLE

Technology is not something that humankind can control.
It is an event that has befallen the world.

—John Gray, *Straw Dogs*

Jordon Peele's 2022 movie *Nope* is many things—a modern take on the classic monster movie, a study of spectacle, a comment on our relationship with nature, a reminder of the erasure of the Black presence in cinema. It's also a critique of our media-saturated society. My interest in media drew me in to that aspect of the film over any of the others.[1] The main characters are Hollywood horse wranglers, descendants of Alistair E. Haywood, the Black jockey in Eadweard Muybridge's 1878 series of photographs of a horse running, one of the first motion pictures ever recorded.[2] The technological mediation of experience is less of a theme of the movie and more of a condition of the environment it's set in.

I first started thinking about technological mediation in high school. A friend of mine rented a video camera one summer, and we made goofy videos with it. In one of the shots on one of those VHS tapes, there's a close-up view of a photocopied picture of me (in a Malcolm X pose) I had pinned on the bulletin board in my bedroom. Watching that tape on my parents' VCR, I remember thinking about the layers in that image. It was grainy playback on an old cathode ray tube television of a magnetic tape recording of video imagery of a manipulated photocopy of a photograph of me imitating one of my heroes. Thinking through the layers of that mediation stuck with me.[3]

There's a scene in *Nope* in which former child star Ricky "Jupe" Park (Steven Yuen) is explaining to Emerald (Keke Palmer) and O. J. Haywood (Daniel Kaluuya) what happened during the "Gordy's Birthday" episode of the hit 1990s sitcom *Gordy's Home*. He describes the episode as a "spectacle" that the network tried to bury but that the public was "obsessed" with. One of the chimpanzees playing Gordy, fed up and finally set off by birthday balloons popping, ravaged the set, killed most of the cast, and maimed his costar. Shown intercut with clips from the massacre, Jupe tells the story not by recounting his experience hiding under the table during the harrowing scene but instead by recounting a *Saturday Night Live* sketch in which Chris Kattan plays the rampaging Gordy. Though he describes Kattan as a "force of nature," we never actually see the fictitious *SNL* sketch. We're watching Yuen play Jupe in describing a sketch spoofing a sitcom in a movie. Foreshadowing aside, the layers in this scene struck me in the same way the photocopied image described above did: one event filtered through one medium after another before it is experienced. With all of our screens and things, we take such layers for granted. Ironically, this is one of the scenes in the movie least laden with the apparatus of film.

In Andrew Patterson's 2019 movie *The Vast of Night*, phone operator Fay Crocker (Sierra McCormick) and radio DJ Everett Sloan (Jake Horowitz) come across an errant frequency on the airwaves that turns out to be alien in origin. Like Peele, Patterson layered the movie with references and homage to other science fiction classics. The call sign for the radio station Everett works at is WOTW, which subtly references *The War of the Worlds*, the classic H. G. Wells novel that became the classic Orson Welles radio play. There is a *Twilight Zone* motif that runs through the story via an introductory voiceover and intermittent TV screens. The tiny New Mexico town the story is set in is called Cayuga, which was the name of Rod Serling's production company.[4] As *The Vast of Night* unfolds, telecommunication and recording devices abound: Fay's new tape recorder, the telephone switchboard, the radio station, old magnetic tapes, and an old camera, among others. The presence of the aliens is discovered through some of these, and others are used in failed efforts to capture evidence of their visit.[5]

Recording devices are rife in *Nope* as well, mostly those related to filmmaking, capturing action in motion. Cameras are everywhere: movie cameras, security cameras, cell phone cameras, digital cameras, and even an old flash camera in the bottom of a well. The eye/mouth of the "unidentified aerial phenomenon" even looks like a camera, as does the helmet worn by the TMZ motorcycle man who shows up during the climactic sequence of the movie but is quickly vanquished by the alien, all the while begging O. J. to get his camera and save his footage. Though the rest of the characters here are trying to capture and capitalize on the spectacle at hand, this motorcycle guy represents the pure tourism of the tabloid, the clickbait culture much of the internet has devolved into. He embodies Nahum 3:6, the Bible verse that opens the movie: "I will pelt you with filth, I will treat you with contempt and make you a spectacle." Those last two lines could have been Muybridge talking to his jockey on set.

In addition to lenses, screens on which to view the images captured by the cameras and windows through which to see the outside world are plentiful. Angel (Brandon Perea) announces the arrival of the alien by saying "It's *heeeeere,*" an obvious reference to the arrival of the ghosts via the TV screen in Tobe Hooper's *Poltergeist* (1982). Angel is only a part of the crew because he helped set up the cameras and continues to monitor them against O. J. and Em's wishes. It's a whole world seen through lenses, windows, and screens. Jupe even refers to the aliens as "the Viewers."[6]

Em might have gotten the money shot at the end of alien battle, but it meant less to her than seeing that O. J. has survived. The latent lesson of *Nope* is that as much as we might want to mediate the moment, to capture the spectacle we experience, we can't take the feeling with us. The medium is the bad miracle. "My experience is what I agree to attend to," wrote the philosopher and psychologist William James.[7] Most of life's great scenes happen outside of lenses and screens, uncaptured by media of any kind.

ACTUAL SIZE

For a list of all the ways technology has failed to
improve the quality of life, please press 3.
—Alice Kahn

I journal like a lovelorn kid in middle school, and I've been keeping a journal almost since I *was* a lovelorn kid in middle school. In my senior year of high school, I met a girl. I started writing poems about her on receipts, handbills, and other various scraps of paper. My writing about her was so prolific that I started keeping it all in a notebook. I've been keeping such a notebook ever since. Around the same time, I started keeping a day-to-day journal as an extension of the poems. I've kept some form of both off and on ever since.

For me, journals are like asides that begin and never end, parentheticals or paratexts running on in the margins of other projects. Though the writing and thinking there ends up in other pieces that are crafted for consumption, the content of the journals themselves is for me only. Mine are full of drawings, diagrams, lists, and quotes from dreams, friends, films, and books. A lot of what follows was seeded in those and the various other journals I've been keeping, each entry a break to record and reflect on recent readings and events.

Similarly, in the late 1970s when John Naisbitt was researching his best-selling 1982 book *Megatrends: Ten New Directions Transforming Our Lives*, he had a filing system of shoeboxes. Each box was labeled according to a major trend he had spotted in local newspapers from across the country and filled with the relevant clippings from those papers.[8] He was parsing patterns he found across all the local news of the nation, archived by hand. His method of research has been rendered obsolete by the all-encompassing internet, and it might sound silly now, especially to the digital natives who have grown up with instant access to everything.[9] Such is the march of media technology.

What I have written here is inherently informed by a similarly generational point of view. I was born in December 1970. I got my first computer at age eleven, though I didn't get online until my mid-twenties. I didn't have a cell

The author at age eleven in the *Athens (Ala.) News Courier*.

phone until I was in my mid-thirties or a smartphone until my mid-forties. I haven't had a car since the 1990s, and as an adult I have never owned a television. As much as I chase and study the technology of my time in some respects, I resist it in many others. The concerns addressed in this book reflect all of the above, but none of them is intended to assume yours are the same.[10] Your memories and experiences may vary.

One piece of advice I have often gotten as well as given as a writer is to find a model for what you're writing. One of the many models I used for this book was John Gray's *Straw Dogs: Thoughts on Humans and Other Animals*. In the preface to that great book, Gray writes, "I have made rather extensive use of quotations—not, I believe, in order to lend authority to an unfamiliar way of thinking, but simply to illustrate what it might mean."[11] I love this sentiment, not only in form but also in content. So much of this material is offered not to lend authority but to illustrate what it might mean to do so. As I attempted to formulate an approach to studying the layers of mediation above, I found others trying to do the same. A few of these people deserve special mention as their ideas are found throughout what follows.

Released in 1979, Douglas Hofstadter's first book, the Pulitzer Prize–winning *Gödel, Escher, Bach: An Eternal Golden Braid*, is an expansive volume that explores how living things come to be from nonliving things. It's about self-reference and emergence and creation and lots of other things. For the cover of his heady tome, Hofstadter carved two wood-block objects such that their shadows would cast the book's initials when lit against a flat backdrop. He went the extra step of working in the initials for the subtitle as well. If I were to follow through with *The Medium Picture*'s cover image, obviously inspired by Hofstadter's, and name this book after the three people who shaped it, it would be called *Gibson, Eno, McLuhan*.[12]

William Gibson's thought hangs heavy over all of these pages. Nearly every time I came up with a way to theorize and think about the process of technological mediation, I found that Gibson had gotten there first. His novels and essays are subtle sources of the nuances of the now, a wealth of ways to think about how we currently live with our world. I have quoted him and lifted from his work liberally, and he was kind enough to blurb the results.

Brian Eno crept into these pages more than I expected as well. His thinking on music and media is both vast and deep, but one idea in particular, Eno's idea of *edge culture*, outlined briefly in his 1996 memoir, *A Year with Swollen Appendices*, became the start for several of the threads in this book. He kindly granted me permission to explore and expand on it further. I can only hope I've done his thought and gesture justice herein.

"Call it religion or call it optimism," Marshall McLuhan biographer Douglas Coupland writes, "but hope, for Marshall, lay in the fact that humans are social creatures first, and that our ability to express intelligence and build civilizations stems from our inherent social needs as individuals."[13] Anyone who tries to do what I have attempted here has to contend with Marshall McLuhan, whose name and thought will recur throughout these pages more than any other. He held a deep disdain for the media and its attendant technology. Despite his insight, foresight, and prescience, he *hated* this stuff.[14] McLuhan wouldn't have liked our current reliance on technology and connectivity one bit, but he would've found it interesting.[15] His son Eric McLuhan also makes several appearances, and his grandson Andrew McLuhan provided the foreword. There's no overstating the pride I feel in having my work this close to their legacy.

My own interests and experiences have informed this book more than any I have written in the past. Along with media theory in the traditions of media ecology and media archaeology, I have written about my lifelong passion for skateboarding and its media. Using what Howard Bloom calls *passion points*, I have included imprinting moments from my life as a music fan, approaching media technology through music recording and its playback. The technologies involved in commodifying and consuming music have a rich history, and I leverage a brief version of it here. As Greil Marcus wrote of his own book *Lipstick Traces: A Secret History of the 20th Century*, "This book does not pretend to be a history of any of the movements it addresses."[16] Likewise with this present volume. In addition, I have used some of my favorite artists as examples, including the bands Fugazi, Radiohead, and Gang of Four, and the artists Laurie Anderson, Christian Marclay, John Cage, and Richard Long.

What follows are three sections of three chapters each. The first focuses on the separation caused by our media technologies. That is, the ways we put them between us and our selves, us and each other, and us and our world. The second explores the thresholds that result from our technologically mediated world, and the third looks ahead, reconciling the two. If we shape our tools and thereafter they shape us, then this is the shape of us to come.[17]

PART I
SEPARATION

Forking Paths

Technology is a way of organizing the universe so
that man doesn't have to experience it.
—Max Frisch, *Homo Faber*

Joshua Joseph has no great hatred of modern technology—
he just mistrusts the effortless, textureless surfaces
and the ease with which it trains you to do things in
the way most convenient to the machine.
—from Nick Harkaway's *Angelmaker*

Every day, computers are making people easier to use.
—David Temkin, *In-Formation Magazine*

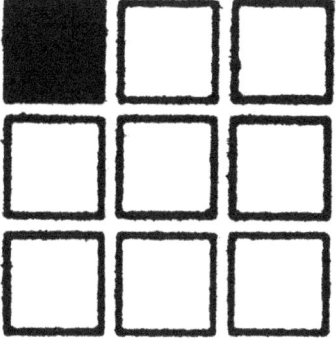

Err Apparent

The subtle pleasure of arriving early and gauging, by the empty period which separates us from the precise time, what we are before we are there. But those who arrive late are doubtless also lingering over an equally perverse pleasure, having taken the time not to be there before they are.

—Jean Baudrillard, *Fragments*

The mass media are genealogical because, in them, every new invention sets off a chain reaction of new inventions, produces a sort of common language. They have no memory because, when the chain of imitations has been produced, no one can remember who started it, and the head of the clan is confused with the latest great grandson.

—Umberto Eco, *Travels in Hyperreality*

The New does not emerge. It erupts, then fades away. It always begins with moments of undefinedness.

—Geert Lovink, *Dark Fiber*

NEW LEAVES

We have participated unknowingly in the creation of a
spurious reality, and then we have obligingly fed it to ourselves.
—Philip K. Dick, "How to Build a Universe That Doesn't Fall Apart Two Days Later"

At the beginning of every story, there's that phase when you feel a little disoriented: the first pages of a novel, the first scenes of a movie or play, the first notes of a song, the first song on a record, the part of the performance where the audience members are still finding their seats. Gérard Genette calls the textual peripheries of books—forewords, prefaces, and so on—"paratexts": "More than a boundary or a sealed barrier," he writes of such introductory material, "the paratext is, rather, a *threshold*, or—a word Borges used apropos of a preface—a 'vestibule' that offers the world at large the possibility of either stepping inside or turning back. It is an 'undefined zone' between the inside and the outside."[1] A beginning is a bifurcation, a deviation that leads off from the current path. Everything starts with a choice between one direction or another. A decision regarding which way to go marks the beginning of possibility while foreclosing another. One possibility is sacrificed so that another may be pursued. The first steps beyond that point take you into alien territory.

Each time we move to a new city we experience the sensation described above over and over as the city slowly takes shape in our minds. Every new place we try to locate (e.g., the closest grocery store, the post office, and rendezvous points with friends) is a new exercise in orientation. "There is no way to relive that moment when you first walk into a new city and it is really fresh," says the filmmaker Richard Kelly. "Your take on the geography, your take on the environment is completely new—there is something very exciting in that."[2] These are experiences you can never reexperience. Once you are familiar, they're gone.

For example, I have lived in Seattle four separate times since first moving there in 1993, and while I have my memory tested with each return after being away, that initial experience of orienting myself is gone forever. To be sure, things seem familiar in an unfamiliar way every time I return. In 2006 I returned to my high school for the first time in seventeen years. As I sat in the auditorium, watching the percussion band play—the same auditorium I'd sat in so many times before—I had a distinct dreamlike feeling of unfamiliar familiarity. It was like nothing I'd experienced before aside from the most acute sensations of déjà vu.

To put it another way, have you ever seen a record you've already heard a million times and suddenly wanted to buy it again because you crave the feeling it gave you the first time you heard it? Yet you know you that while you might be able to recapture something of that long-lost feeling, you'll never get the full effect again. There is an area of the brain, the amygdala, that is said to be the main controller of emotional memory. This is where these initial experiences are archived in your head, and it's where things change after that initial orientation. You can't change them back. That's the entropy of experience.

The mental space just before we feel oriented and something feels familiar is the discernment of difference, a threshold of perception. In psychophysics, the study of relationships between physical and mental phenomena, the smallest measurable unit of such a perception is called the "just noticeable difference," or the JND.[3] Hearing is a product of this difference. Sound is nothing more than vibrations among other vibrations. A constant sound can seem like a kind of silence, receding into the background of your consciousness until it stops abruptly, leaving true silence, or until it is accompanied by a louder sound, splitting the heard world in two. Hearing, in a cognitive sense, is less about being able to detect sounds than it is being able to tell the difference between these vibrations.[4] Any new experience is the same phenomenon: noticing a difference, noticing differences.

In the last one hundred years, there have been two major shifts in media technology: the separation of communication from transportation, starting with the telegraph and continuing to radio and other broadcast media and to the telephone and the internet, and the move of representation from the page and the stage to the screen.[5] Both of these shifts—the invisibility of delivery networks and the ubiquity of screens—established not only a media infrastructure but also a whole new media environment, starting with the rapid adoption of television.

YOU'RE SOAKING IN IT

Nobody seemed restless, this after all being the Mellow Sixties, a slower-moving time, predigital, not yet so cut into pieces, not even by television.
—from Thomas Pynchon's *Vineland*

Between 1948 and 1955, television invaded almost two-thirds of American homes.[6] Movie ticket sales plummeted over the same time.[7] In less than a decade, TV became the nexus of media presence and personal life. As Marshall McLuhan, the prophet of the television age, famously pointed out, its changes were large enough in scale to be considered environmental. Seeing the diffusion of these technologies as ecological highlights the nuanced changes each of them brings on. Medium theorists Harold Innis, Marshall McLuhan, Walter Ong, and Neil Postman are widely considered the forebears of the study of media as ecology.[8] "*Media Ecology* is a term I invented when we were at Fordham," Marshall's son Eric McLuhan said in 2010. "I discussed it with Postman, and he ran with it."[9] Postman described media ecology as concerned with "how media of communication affect human perception, understanding, feeling, and value; and how our interaction with media facilitates or impedes our chances of survival. . . . Media ecology is the study of media as environments."[10] As a new subdiscipline, media ecology has blurry peripheries. The study of media and communication are central, but the field remains open to the history and philosophy of technology, music, literature, and cultural studies.[11]

Marshall McLuhan and Walter Ong both studied the epochal transition of their time from oral to print culture and how it changed more than just spoken and written communication.[12] They were also able to witness the changes that radio, telephone, and television technologies wrought. McLuhan tackled the changes in scale that new technologies exacted.[13] He looked at every media technology in relation to every other, which is a key tenet of media ecology, that the interrelationship of people, technology, communication, information, and media makes up an ecology.[14] Just as it does in a natural ecology, when one thing changes (e.g., the climate, the flora, the population of predators or prey), the interrelationship changes. According to Postman, the central question that McLuhan posed was "Does the form of any medium of communication affect our social relations, our political ideas, or psychic habits, and of course, as he always emphasized, our sensorium?"[15] That is why, as we will see, the Walkman changed our relationship to the world by adding an individualized soundtrack to our experiences but also changed our relationship to music, our environment, our devices, and each other.

As Ong once put it, "Today, it appears, we live in a culture or in cultures very much drawn to openness and in particular to open-system models for conceptual representations. This openness can be connected with our new kind of orality, the secondary orality of our electronic age."[16] "Secondary orality" reminds one of the original names of certain technologies (e.g., "horseless carriage," "cordless phone," and "wireless technology"), as if the real name for the thing was yet to come along. If Ong's ideas seem similar to McLuhan's, it should be noted that the latter supervised the former's master's thesis, both pushed for a more systemic view of media technology, and they shared many concerns about the onset of electronic media and the changes it brought.

Technological change is not merely additive, though. It changes everything. Ong wrote, "The advent of newer media alters the meaning and relevance of the older. Media overlap, or, as Marshall McLuhan has put it, move through one another as do galaxies of stars, each maintaining its own basic integrity but also bearing the marks of the encounter ever after."[17] Old media technologies cast new ones in a new light, and vice versa, providing glimpses of the nature of the water in which we swim.[18] The changes are the thing.

Like McLuhan and Ong, Friedrich Kittler studied the technological epochs of his time. The changes in media and the ramifications of those changes were his main concerns. Kittler was especially interested in inscription technologies. He concerned himself with the ways that the advent of gramophone, the typewriter, and film altered the place of writing's dominance on our culture. He viewed language as dispersed via these new technologies and transmitted via media more specific to their content (i.e., via images, sound, etc.). They broke up the "monopoly on the storage of serial data" previously held by the book.[19] The interrelationship of these technologies he termed *discourse networks*, arguing that they "designate the network of technologies and institutions that allow a given culture to select, store, and process relevant data."[20] "In a literal sense," N. Katherine Hayles writes, "technologies of inscription are media when they are perceived as mediating, inserting themselves into the chain of textual production." Invoking John Culkin's description of McLuhan's work, Kittler acknowledged that we shape our tools, and they shape us. He repeatedly referenced Friedrich Nietzsche's observation that "our writing tools are also working on our thoughts."[21] Nietzsche was writing about his typewriter, but the sentiment is easily more broadly applicable to the massive changes in media technology since.

Each new medium not only changes but also reconfigures the current ones. For example, the onset of electronic media extended the reach of literacy by reinforcing the use of writing and print media.[22] No one medium or technology stands alone. They must be considered in concert. Moreover, according to the approach of media ecology, to be literate in the all-at-once world of electronic media is to understand its systemic nature, the inherent interrelationship and interconnectedness of technology and media.[23]

Another relevant aspect of the ecological view is that of the biases created by different forms of media. Harold Innis stated that any new medium possesses the tendency to distort communication by emphasizing either time or space and by deemphasizing the other. Sometimes the temporal factor matters, and sometimes it doesn't.[24] McLuhan argued that the biases of media are due to their distortion of the ratio between our senses, as different media appeal to and emphasize the use of sight, hearing, touch, and so on.[25] Referring to images, he wrote, "These forms provide a world of inclusive gesture and dramatic posture that necessarily is omitted in the written word."[26] Their biases certainly don't end there, though.

Each new technology frees us from something. Writing freed language from speech. The printing press freed language from writing. The telegraph freed information from transportation. The telephone freed real-time communication from physical proximity.[27] The cellular phone freed communication from location altogether. Within each of these newfound freedoms, a paradox exists: each liberatory advance tethers us to a new form. Even as technology frees us from something, it binds us to itself. If we take this technological binding to its logical extreme, we can view our contrivances as enslaving us.[28] Brian Eno has a great phrase for our entering this relationship: *a surrender with risk.*[29] Our feet become shoe-shaped, sculpted by a lifetime of wearing shoes. Our bodies become car-shaped in the sense that they are sculpted not by walking but by traveling by car. Our eyes are reshaped by staring at various screens.

New technologies are new beginnings. They are the initial conditions from which new forms of our media are born. Feared and disparaged at first, technological advances are eventually welcomed into the world. They change our minds. They change our relationship with our worlds and with each other. They produce noticeable differences.

Moreover, each new contrivance changes not only what is expressed but also what can be expressed.[30] Technology curates culture. Not unlike learning new words, every advance is an addition to our media lexicon. Our media vocabulary includes those technologies with which we feel facile and familiar. What happens when we are fluent in technology's tongue? What happens when technology infiltrates every aspect of our existence?

SEPARATION ANXIETY

We keep waiting for the robots to crush us from the sky.
They sneak in through our fingertips and bleed our fingers dry.
—Milemarker, "Frigid Forms Sell You Warmth"

"The voice inside your head has become a different voice," writes Douglas Coupland. "It used to be 'you.' Now your voice is that of a perpetual nomad drifting along a melting landscape, living day to day, expecting everything and nothing."[31] Fears of technology's takeover are nothing new. Anxieties about the merging of humans and machines have been prevalent since humans and machines have coexisted. From Thomas Edison's first recorded sound in 1877 to the arrival of the personal computer, which has since crept into every aspect of our lives, and especially more recently since personal media have infiltrated our existences, shrinking attention spans, that worry has only increased.[32] From artificial intelligence and humanoid robots to microchip implants and the transhumanist dream of uploading consciousness, the melding of biology and technology has become both an everyday occurrence and more of a platitude than a prediction.

As soon as humankind started externalizing its knowledge through the technology of language, we began blurring the lines between our tools and ourselves. Toward the end of the first decade of the twentieth century, we reached a watershed: we created more data in 2009 than we had in all the years before combined.[33] The average American family now spends more every month on communication technologies and connectivity than it does on food.[34] We also watched more television in 2009 than ever before, and that's not even counting time spent online, where, while surfing websites, we are supposedly interacting with our media rather than passively viewing it.[35] Yet there we're adrift in the endless update culture of social media. If it wasn't posted today—or in the last few hours—it disappears into irrelevance. If it's too long, it doesn't get read at all. These are not rivers or streams, they're puddles. All comments, references, and messages, no story. The personal narrative is lost. It's the age of "tl;dr" (too long; didn't read)—the twenty-four-hour news cycle, a present made up of the past, and advertising interrupting everything. We don't surf anymore; we just sit back and watch the waves.

Again, anxieties about technology taking over our lives are not new. An illustration by Harry Grant Dart from a 1911 issue of *Life* magazine depicts an extreme example. A man is shown seated in the middle of a room where speakers, tubes, ducts, projectors, and printouts provide his every need—comfort, nutrition, information, entertainment. He needn't ever leave his chair. An original thought never need pass through his head.

Significantly, 1911 was the year that the American mechanical engineer Frederick W. Taylor's *The Principles of Scientific Management* was published. In Taylor's near future, technological systems would take over human supervisory duties, ushering in an era of modern management.[36] His visions have, in large part, come true. Organizational theory and technology have coevolved so closely that our exposure to technological innovations often happens at work. We become comfortable in our working lives with things like computers and mobile technologies before they infiltrate our homes, often tethering us to work at all hours.

Much later in the twentieth century, artist Jeff Nicholson depicted the main character in his 1994 comic *Through the Habitrails* enduring technology-enabled weekends in which he enjoys all of his leisure activities all at once. His work week leaves him so drained of enjoyment and so far from his interests that he seeks to fill the gap as quickly as possible. "My stimuli is taken directly to my nerve endings and orifices," he says, "and I take it in and in and in with clenched teeth and a fibrillating heartbeat." Attempting to reconnect the separated parts of his life pulled apart by workaday technology, Nicholson's nameless protagonist relies, ironically, on technology. The cognitive scientist Gary Marcus writes, "Forms of entertainment like music, movies, and video games might be thought of as what Steven Pinker calls 'pleasure technologies'—cultural inventions that maximize the responses of our reward system. We enjoy such things not because they propagate our genes or because they conveyed specific advantages to our ancestors, but because they have been *culturally* selected for—precisely to the extent that they manage to tap into loopholes in our preexisting pleasure-seeking machinery."[37]

Even if we don't fully adopt the apparatus, submitting ourselves by strapping the machines directly onto our bodies, we are still swallowed up by them, saturated by their sounds and images. In his book *Last Futures: Nature, Technology, and the End of Architecture*, Douglas Murphy writes, "If any environmental experience could be created by a few slide projectors in a darkened room, what was the point in ever more adventurous architecture? If the superstructure of the future was to be as blank and anonymous as a giant space frame or dome,

"We'll All Be Happy Then," Harry Grant Dart, *Life*, 1911.

the only 'design' required would be that of the individual's multimedia experience within that frame, which would be an entirely new field of work, a kind of ever-more-personalized interior design."[38]

Here are two more examples, of a more quotidian nature. Since 2008 I've been riding a fixed-gear bicycle—a bicycle that has a direct connection between the front and rear gears and the rear wheel. Thus, the pedals and the rear wheel work in concert at all times, so that the bicycle doesn't coast. Given the extra work and hazards associated with such a vehicle, people often ask me why I ride one. What's the appeal? One reason that fixed gears are so rewarding is the direct connection between human and machine and environment. No matter the terrain or conditions, your body is always at work, negotiating the ride. Feeling the seamless interface of the bike as a technical extension of your body is exhilarating. Through the medium of the bike, you are directly connected to the experience of motion and the environment you're moving through.

Another example: Walking to class one day at the University of Texas at Austin, I realized I had the option of taking the elevator up to the seventh floor and then going to the climbing gym afterward. It struck me as odd that the two things were completely disconnected. Getting to a higher floor in one building and the act of climbing up a wall in another were totally disassociated, even though they were essentially the same act. In her book *Wanderlust: A History of Walking*, Rebecca Solnit addresses a similar separation between physical activity and our machine-enabled tasks, writing, "What exactly is the nature of the transformation in which machines now pump our water but we go to other machines to engage in the act of pumping, not for the sake of our bodies, bodies theoretically liberated by machine technology? Has something been lost when the relationship between our muscles and our world vanishes, when the water is managed by one machine and the muscles by another in two unconnected processes?"[39]

We drive cars to the gym to run miles on a treadmill. Inclement weather notwithstanding, why don't we just run down the sidewalk? The activities are disconnected. If our culture is essentially technology-driven, then what kind of culture emerges from such separations between our physical goals and our technologically enabled activities?

Much of the alienation we feel from our technologically mediated lives comes from a separation between physical goals and technology's "help" in easing our workload. It's like taking a shower when you're thirsty.

Robert Rosenberger writes, "When we become accustomed to embodying a technology, it takes on a kind of experiential 'transparency.' . . . Our technologies sometimes greatly change our abilities and our experience. Yet through our design practices, and through the ways we as users become acclimated to our devices, we may in many moments remain barely aware of those changes and, indeed, sometimes barely aware of the device itself."[40]

Though some argue we will only be happy when we're cocooned in a matrix of our own making, our devices are divisive, alienating us from each other and the world around us and even our own bodies.[41]

Jeff Nicholson's remedy for the forty-hour workweek:
the all-at-once weekend. From *Through the Habitrails*.
Reprinted by permission.

There are at least three types of separation at work in contemporary culture: one between ourselves and our environment (e.g., pumping water vs. pumping iron), one between ourselves and each other (e.g., individual distraction vs. global connection), and one between ourselves and our knowledge (e.g., thoughts in our heads vs. words on a page), with technology wedged in between the two terms in all three cases. The latter dynamic overlaps the first two and is one of the paramount issues of our age. Often there isn't a choice regarding what is easier, more convenient, or more fun, much less what is more acceptable. Inevitably, the technology in place makes only one path available. The digital revolution, ballyhooed as ushering in an age of personal empowerment and infinite consumer choices, imposes new limitations.

It's unfortunate that the word *technology* is now almost exclusively associated with computers and other digital devices. Corporate policy makers, educators, and communication scholars debate whether "technology" should be allowed in meetings or the classroom—as if pencils, paper, or, for that matter, language aren't technologies. By the strictest definition, a technology is the knowledge or practice of a skill. Echoing McLuhan and McLuhan above, Solnit's "best definition" is that "a technology is a practice, a technique, or a device for altering the world or the experience of the world."[42] Finding out what exactly it alters and how is part of our purpose here.

One of the narrators in Samuel Butler's 1872 novel *Erewhon* says "Why may not there arise some new phase of mind which shall be as different from all present known phases, as the mind of animals is from that of vegetables?"[43] He goes on to describe how a chicken egg, a chicken's nest, and an egg-cup are all machinery designed for the same task, and he points out how our tools extend our limbs ("a machine is merely a supplementary limb") and calling us "machinate mammals."[44] As if anticipating an inevitable combination, he seems to describe not so much a merging of humans and machines but an entirely new form born of the two. "Who can draw the line?" he contests. "Who can draw any line? Is not everything interwoven with everything? Is not machinery linked with animal life in an infinite variety of ways?"[45] Adopting a new technology represents the *surrender with risk* Brian Eno has talked of, as every new contrivance augments some choices at the expense of others, giving and taking simultaneously.[46] What we lose in that exchange is often unknown to us.

Taking another tack, in his book *The Nature of Technology*, Brian Arthur defines technology in opposition to nature, imagining the opposing forces as "tectonic plates grinding inexorably into each other in one long, slow collision."[47] As often as it has been employed elsewhere, this is a premise of limited promise. Technology is an extension or continuation of nature. They're parts of the same continuum. Viewing them as a false binary, Arthur envisions a world where all our modern technologies disappear, yet paradoxically we're still left with some tools. He writes, "We would still have watermills, and foundries, and oxcarts; and course linens, and hooded cloaks, and sophisticated techniques for building cathedrals. But we would once again be medieval."[48] Compare that to the following passage: "While technological progress AS A WHOLE continually narrows our sphere of freedom, each new technical advance CONSIDERED BY ITSELF appears to be desirable. Electricity, indoor plumbing, rapid long-distance communication . . . how could anyone argue against any of these things, or against any of the innumerable technical advances that have made modern society?"[49]

This is from Theodore Kaczynski's manifesto. Kaczynski, better known as "the Unabomber," spent decades targeting what he saw as the evils of technological advance by murdering its creators. He continues, "Yet . . . all these technological advances taken together have created a world in which the average man's fate is no longer in his own hands or in the hands of his neighbors and friends, but in those of politicians, corporation executives, and remote, anonymous technicians and bureaucrats whom he as an individual has no power to influence."[50]

Drawing such an arbitrary line in the sands of time is the mistake that technophobes make.[51] As if language and writing aren't technology. As if harnessing fire or clothing ourselves aren't technology. Our current existence is the result of a symbiotic relationship with our technology.[52] Not that Arthur and Kaczynski are against technology, but making distinctions as such is not only treacherous, it's ludicrous. As William Gibson points out in Mark Neale's 2000 documentary, *No Maps for These Territories*,

> I think what I'm most aware of is the extent to which people are unaware of the extent to which they've been interpenetrated and co-opted by their technology. And I take it for granted that I've been. . . . I think a lot of people today have as this sort of a Rousseau-esque idea that it's possible for humans to return to "the natural state." But . . . I think it's not, and, if it were, they really wouldn't like it. I mean, I'm immune to a number of really, really terrible diseases because I was inoculated against them as a child. That's technology. I'm a male human in my fifties, and I still have most of my teeth. That's tech-

nology. I'm myopic, to the point of near-blindness, and yet I can see. And that's technology. It's too close to us to be very aware of it. If we could be stripped of it—which we can't be, because it's actually altered our physical being—we'd be pretty unhappy, you know?[53]

Any attempt to return to some previous era or so-called natural state is a futile attempt to get back across the line we've drawn for ourselves, a line that implies a perfect level of technology, a perfect amount of technological mediation.[54] Arthur's is just past watermills, linens, and cathedrals. Kaczynski's is past electricity, indoor plumbing, and long-distance communication. Some would place any of those technologies far down the path to perdition. Kaczynski tried to return to what he saw as a more natural state, living as self-sufficiently as possible in a shack in bucolic Montana. Encroached upon there by the oppressive powers of technological society, he began fighting back with strong words and mail bombs.

Considering our relationship with technology, zooming out to see what is essential and what is desired, reminds me of two lessons. One is from Ray Bradbury's 1953 novel *Fahrenheit 451*, where the character Faber outlines three essential things:

1. Quality, texture of information.
2. Leisure to digest it.
3. The right to carry out actions based on what we learn from the interaction of the first two.[55]

Unfortunately, these seem more germane than ever, as it has gotten more and more difficult to find any of them.

Also, one of the many lessons of chaos theory was that the limits of numerous traditional scientific and mathematical approaches had been reached.[56] The elements filtered out by the methods in use kept edging in, refusing to be ignored. Information theorists, physicists, and mathematicians were all grappling with similar, persistent problems: noise in phone lines, measurements that varied wildly at different scales, fluctuations in computer-generated weather, the onset of turbulence in vastly different dynamical systems—together signaling a new paradigm that Hayles calls "orderly disorder."[57] New lenses were needed to see a more finely grained world. New tools were needed to measure it. New ways of thinking were needed to understand what had changed.

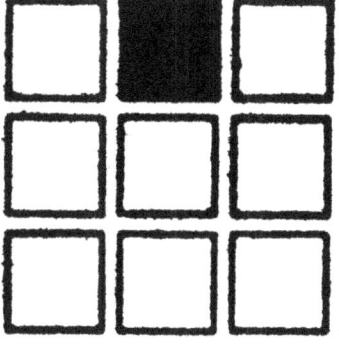

Audible
Arrangements

The choice is not between quick fixes and dull calculation.
This is what liberal education is meant to show them. But as long
as they have the Walkman on, they cannot hear what the great
tradition has to say. And, after its prolonged use, when they
take it off, they find they are deaf.

—Allan Bloom, *The Closing of the American Mind*

Hit the rewind and move and turn the black plate
Get you a tape, watch the crew wreck
Spin the vinyl till it melt and get a phat new cassette.

—The Nonce, "Mixtapes"

The psychogeographer knows that the world cannot
be recorded, it can only be remade."

—Stewart Home, "How I Discovered America"

PATINA TURNER

We are that strange species that constructs artifacts
intended to counter the natural flow of forgetting.
—William Gibson, *Distrust That Particular Flavor*

I was raised by record stores. That's where I was while my mom was shopping. I was in Peaches or Coconuts or Tower or whatever suburban chain was in the town we lived in at the time. When I started, I was barely able to see over the counter. It was a childhood spent digging in the racks, gawking at album covers, and occasionally buying a twelve-by-twelve-inch cellophane-wrapped record to take home, unwrap, and spin.

Releasing the secrets hidden in those grooved plates of vinyl, what Alex Weheliye calls "the allurements that lurk in the crevices," was one of my pre-adolescent joys.[1] My parents were never into music, and I don't have any older siblings, so recorded music was a world I discovered on my own and with my friends through shared enthusiasms and swapped records, tapes, and eventually compact discs. We found music in a network of like minds. Tapes would get copied and passed around after coming in the mail from far-flung friends. Someone would hit a record store after traveling to a big city far away and come back with rarified sounds. Someone else would get a pile of records through a mail-order service. These were the raw materials of the tapes we made and the songs we shared. Little did we know those sharing practices were a moment in time, soon to be wholly transformed by the then-fledgling internet.

I survived my undergraduate years working at record stores. Not surprisingly, the lulls behind the counter were largely spent talking about and sharing music. My coworkers and I would all bring in our shoebox-sized tape and CD cases, each stocked with a dozen or so selections for the shift. This was the early 1990s, a scene somewhere between the "top five" lists and high fives of the 2000 film *High Fidelity* and the shoptalk and behind-the-scenes schemes of the 1995 film *Empire Records*. There were lots of lists and mixtapes. There was a lot of judging and clowning but even more sharing and playing each other new music. Even with our employee discounts, our discretionary budgets back then were small, and cassettes cost around ten dollars, while CDs were closer to seventeen. I remember talk in the music industry at the time promising that the CD would soon cost the same as a tape, hinting at a cheaper CD. Instead, CDs stayed the same, and the tape price eventually edged upward. LPs were all but gone at the time, with fewer and fewer new releases even appearing in the format.

Prohibitive pricing notwithstanding, buying music was always a risk. You might have heard a song or two from a friend or seen a late-night video, but most of what one might buy was unheard, a mystery that could turn out to be quite disappointing. You never knew when you were going to have enough money to buy another record, so if you had money for one in the first place, you had to hope whatever you were buying was good. If it was good, it was your whole life until you had money to buy another one.

Where we have access to almost any artist's complete catalog online now, your relationship to the music you have to look for is different. "For those of us who were pre-internet and post-internet, we can really see the distinction," Dischord Records cofounder Ian MacKaye tells me. "I'm not a Luddite, and I'm not nostalgic. I don't care about any of that. But the reality is that the relationship I had with music at a time where I would only be able to afford one or two records, and I would just have to go and listen to that record until I [saved] up for the next record. I would listen to one record, you know, forty times in a row. That experience is much more difficult when you have four million musical choices at your fingertips."[2]

Then as now, retail space in record stores is precious. Product must move. If it's not moving, it gets extra incentive to do so. This means bargain bins. Typically located close to the front of the store, these displays were piled with releases the record labels were unloading at a discount or whatever stock the store needed to get rid of. A tape or LP at a fraction of the suggested retail price was too good to pass up. Subsequently I found many lifelong loves in those racks. Staples-to-be like Naked Raygun, the Jesus and Mary Chain, and Gang of Four were joined by lesser-known acts like Abecedarians, the Devlins, and Bleached Black. These were complemented by cassette dubs, often cribbed from the mixtapes from friends near and far, and the occasional full-price splurge on a new release. It was a weird blend of sounds, idiosyncratic in its breadth and fickle in its focus. Behind the counter at the record store, whatever tapes or CDs you had in your case were to be debated and defended until closing time.

When I first got an iPod in 2003, I thought that social practice would continue. Around the time that I procured my refurbished player, a friend of mine came to San Diego to visit. One day he was hanging out with some of his old college friends, one of whom had a new boyfriend. My friend snagged the dude's iPod from her and was judging her new beau on the merits of his MP3s. Maybe this has happened more often than I'm aware, but in my experience it was a rarity. With the coming of the digital age that rendered records and CDs obsolete, and with their demise much of the communal connoisseurship of music fandom, our individual listening experiences tend to be as insular as the devices that facilitate them.

Studies of the digital sharing of music call it "playlistism," a subcultural ritual that reinforces the links between music and collective identity through the practice of sharing playlists. Assuming that we compile playlists to represent our identities, the sharing of them should show how we present ourselves through music. Citing a 2001 study by Brown, Sellen, and Geelhoed, Mariya Valcheva found that sharing via peer-to-peer networks "confounded the traditional way of possessing and sharing music . . . thus instigating a shift, on one hand, towards a citizen/leech styled community where music sharing interaction tends to be anonymized."[3] We don't use P2P spaces to share in the traditional sense. In contrast, "playlistism is underpinned by the practice of capturing and contributing one's 'music personality' in the form of playlists that are either published online or shared through portable devices."[4] As Christiane Gelitz puts it, "We are what we like." In compiling and sharing our musical tastes, we go from saying "I like that" to "I'm like that."[5] Playlists have overcome not only the constraints of mixtapes but also radio and even records for a lot of music fans, but sometimes we need something to dampen our desires.

LIMIT TO WIN IT

Good art is produced under strict limits. Forces you to work with what you've got, to focus, right? There was no focus at the end. No control or vision.
—Mike, in Tim Maughan's *Infinite Detail*

Sounds are woven with memory.
—David Toop, *Haunted Weather*

On a continuum of identity or intimacy, mixtapes are closest to our hearts, followed by playlists, and then open sharing via peer-to-peer networks and streaming services. Intimacy requires restraint. "With tools, we crave intimacy," Brian Eno writes. "This appetite for emotional resonance explains why users—when given a choice—prefer deep rapport over endless options. You can't have a relationship with a device whose limits are unknown to you, because without limits it keeps becoming something else."[6] A mixtape has well-known, predetermined limits. The physical limits of cassettes make compiling them more intimate, more personal. A sixty-minute cassette is far more personal than an endless playlist, because, unlike playlists, cassette tapes have preset limits, and confronting and conforming to these limits is a sign of skill and care. As one mix-taper puts it, "The creator must not only choose songs that go well together, but he [also] can't just decide to end it after 30 minutes. He has to fill both sides to their predetermined length, with as little blank space at the end of each side as possible. A mix tape is a work of art."[7] Given the added labor involved in compiling a cassette versus putting songs on a playlist, mixtapes are more valuable to us. We cherish such amateur media not in spite of their technical shortcomings but because of them.[8]

The very existence of the DJ, a "disc jockey," a most antiquated designation, is due to the technical shortcomings of musical playback devices. As party motivation, reel-to-reel tape decks, record players that allow one to stack records, and multidisc CD-changers are sloppy at best. Moving a crowd for more than the length of a song, a CD, or one side of a record, requires ingenuity. It requires consciousness, attention, intuition. It requires a human at the helm.

The stand-alone MP3 file recalls the scattered nature of seven-inch singles from the punk-rock era. "I felt I needed to hear these records in a more time-fluid way," Sonic Youth's Thurston Moore says, "and it hit me that I could make a killer mix tape of all the best songs from these records."[9] Split from their original format, the songs take on new life in a new sequence of songs. With single songs available for cheap, ownership of the medium shifted further from the artist and industry to the individual listener to record and arrange as they liked.[10] The cassette tape is still the original site of the shift from home recording to personal drift, as well as the societal switch from the corporation's copyright to the individual's right to copy. "The tape cassette is a liberating force," Malcolm McLaren proclaimed in the early 1980s, "which is why the record companies are so petrified."[11] William Gibson adds, "The tape recorder was the first widely available instrument that allows you to manipulate media."[12] Long before CD burners became standard equipment on the personal computer and MP3s and streaming liberated music from physical formats altogether, cassettes made recording and customization possible. We

take it for granted now that we can curate, consume, and manipulate music, media, and information as we see fit.

"Home taping is killing music." It sounds funny now, but the British Phonographic Industry—BPI, the English sister of the Recording Industry Association of America (RIAA)—was incensed. Their attitude was that every blank tape sold was a record stolen. "BPI says that home taping costs the industry £228 billion a year in lost revenue," McLaren said in 1979, "so they're not happy that Bow Wow Wow have already reached No. 25 on the singles chart."[13] The home-taping controversy was custom-made for McLaren. He was managing Bow Wow Wow who had a hit with a song celebrating home taping called "C-30! C-60! C-90! GO!" The numbers refer to the standard lengths of blank cassettes: thirty-, sixty-, and ninety-minute formats. "In fact," added McLaren, "it's the classic story of the 80s. It's about a girl who finds it cheaper and easier to tape her favorite discs off the radio."[14] Home taping made everyone a pirate and a producer.[15] In his "mixtape manifesto," Jared Ball writes, "Mixtapes, as community-based, localized mechanisms of distribution, pose a threat to the process of managing the flow of ideas."[16] The formalized mixtape, in McLaren's conception, "allowed for the kinds of communication ultimately threatening to power."[17]

The cultural vocabulary of these media also includes the slang of their flaws, the vernacular of their limitations. Eno continues, "Even the 'weaknesses' or the limits of these tools become a part of the vocabulary of culture."[18] Needles crackle in record grooves, tapes hiss and whine, and both degrade with wear. These signs of wear evoke a sense of time, so much so that the crackling of vinyl is now simulated digitally as "the phonograph effect."[19] Eno adds, "Whatever you now find weird, ugly, uncomfortable and nasty about a new medium will surely become its signature. CD distortion, the jitteriness of digital video, the crap sound of 8-bit—all these will be cherished and emulated as soon as they can be avoided."[20] These familiar flaws connect us to the media even as they critique the media.[21] This patina, the signs of use, the wear, the evidence of human contact and interaction, are internalized in the cultural language of digital media.[22] The wear is in us. It's cognitive. It is true that digital information is infinitely repeatable with no signs of decay, but that doesn't mean that its significance is inexhaustible.[23]

For example, a skeuomorph is a design element that remains only as an allusion to a previous form, like the phonograph effect, woodgrain wallpaper, the desktop metaphor, or even the digital "page." It's obsolete except in signifying what it supplants. The literary critic N. Katherine Hayles describes the concept, writing, "It calls into play a psychodynamic that finds the new more acceptable when it recalls the old that it is in the process of displacing and finds the traditional more comfortable when it is presented in a context that reminds us we can escape from it into the new."[24] Like euphemisms, skeuomorphs mediate uncomfortable shifts toward an uncertain future, translating the unknown into the terms of the known.[25] We love to recognize a previous medium artifacting into a new one. Some of us more than others.

DIGGING IN THE DIRT

The essence of culture is found in all its artifacts.
—Pete Robinson, in Donald Antrim's *Elect Mr. Robinson for a Better World*

Steampunk, so named to contrast it with the dystopic, digital futures of cyberpunk, represents an excitingly innovative yet alienatingly weird subculture. The genre possesses hints of nostalgia, punk-rock attitude, and a love for self-styled, homemade gadgets. William Gibson and Bruce Sterling provide an easy touchstone with their 1990 book *The Difference Engine*, a revisionist history of the world in which Charles Babbage finished his steam-powered calculating machine and the information age preceded the Industrial Revolution. Sterling described it in 2022 as "halfway between old-school artificial intelligence and today's AI."[26] As if prompted by today's so-called "generative

AI," the novel is written by the machine in the story. That is, the Difference Engine is the narrator.

Terry Gilliam's 1985 movie *Brazil* is another great example. The society depicted is industrial in its aesthetics and institutional in its metaphysics, bureaucratic beyond all hope. Cowriter and director Gilliam's original title for the project was *1984½*, in homage to both George Orwell's novel *Nineteen Eighty-Four* and Federico Fellini's film *8½*, which share the concerns of the mechanization of modern life.[27] "Some of the technology was a deliberate mixture of the futuristic and the Victorian," says Gilliam. "There's a lot of growing up in America in the forties: progress and the utopian vision were always there, with technology as the answer to all of our needs."[28] The movie's atmosphere and ontology operate atop a machine metaphor.[29] Unexpected breakdowns, routine repairs, and replacing parts—hydraulic hoses, pneumatic tubes, and other analog accoutrements—make up a lot of the concerns of the characters in *Brazil*. Gilliam adds, "There was a collective dream of progress and a centralized belief in the perfectibility of man—either through technology or fascism, which effectively amounted to the same thing."[30] Unlike the panopticon of *Nineteen Eighty-Four*, *Brazil* focuses further on the desire for distraction, to escape from such technological fascism through fantasy.[31] The film takes its surveillance society as a given and tries to find a way out through human imagination.

Though Gilliam was satirizing this mechanical society and its supposed utopian aims, early computer inventor Charles Babbage aimed at divining a version of its universal machine or at least harnessing its hidden power.[32] So it goes with steampunk as a whole, as ironic as it is empowering. By focusing on what's behind the page or screen rather than what is on it, whether it is new machinery or new media, media archaeology invites us to question the newness of what we consider new.[33] In his book *What Is Media Archaeology?*, Jussi Parikka introduces the field under the guise of applying steampunk to media studies.[34] Steampunk looks to the past as well as the future and wonders whether certain initial conditions could change the outcome of our mechanical media madness. Digging up pieces of the past, media archaeology seeks the same. The painter Mark Tansey points out:

> Given that the painted picture is a declassified medium (in Marshall McLuhan's sense—a medium that is no longer the dominant conduit or voice of power, unlike television or film) it can take on new functions. One of these

can be as analogue to other representational media in understanding the limits and sensitivities of one as it relates to those of another. We can use the painted picture as a way of studying its own modes of references, its ranges of sensitivity and insensitivity, its deceptions, by way of offering insights into the analogous functions of, for example, film, photography, and television.[35]

In many ways, media archaeology offers an alternative to simply historical views of media. It is "first and foremost a methodology," the media theorist Geert Lovink writes, "a hermeneutic reading of the 'new' against the grain of the past, rather than telling the histories of technologies from past to present."[36] Also consider the indexing of dead media. Media are "dead" based on their manufacture, adoption, business viability, or lacks thereof, but all these aspects vary, overlap, and waver in and out of relevance. Newspapers, radio, television, and telephones, each touted to kill the others, are all still around in some form.[37] From the onset of the digital and imaginary media to dead devices and the world of sound, finding these lineages as such and predicting the present is what media archaeology is all about.[38] Moreover, Parikka points out that media do not really die but rather remain undead. Their materiality never really disappears, leaving behind waste and unwanted residue. Their purposes are renewed, bringing forth zombie media and alternate futures.[39] What if Charles Babbage had finished the Difference Engine? What if one cog in the universal machine were different? What happens when dead media come back to life? Their bones might be buried, but those hisses and scratches hint at a rich past just under the dirt of the digital.

WALK THIS WAY

So we might suggest that the apparent vacuity of the Walkman opens up the prospect of a passage in which we might discover . . . those other cities that exist inside the city. There we move along invisible grids where emotional energies and the imaginary flow, and where the continual slippage of sense maintains the promise of meaning.

—Iain Chambers, *Migrancy, Culture, Identity*

After walking around New York City one day in the late 1970s, Akio Morita, sort of the Steve Jobs of Sony, decided he wanted a personal tape player, one that he could easily carry around and thereby listen to audio tapes wherever, whenever. So impressed was he by his engineers' first prototype that he wanted to take the device to market. One of the products that prefigured the Walkman, Sony's Pressman, was designed for recording as well as playback. Some Sony executives didn't believe the public would want such a device exclusively for playback, much less one that forced them to wear headphones, which seemed strange at the time.[40] Even so, Morita pushed the idea through, and the Walkman was unleashed on the world on July 1, 1979. The rest is well-recorded history.

The personal-stereo story starts before that though, with a lesser-known invention dubbed "the Stereobelt" by its German-Brazilian inventor Andreas Pavel. In the late 1960s in São Paulo, Pavel hosted record-listening parties similar to my experiences behind the counter at the record store. He and his friends mused over philosophy, politics, music, and how they might be able take their music with them around town. Inspired by their discussions, he invented the first portable personal stereo, cobbled together from the lightest headphones he could find and other odd pieces of bulky but wearable audio technology. He saw the device as a way to "add a soundtrack to real life."

After shopping the Stereobelt to several big audio equipment companies to no avail, Pavel filed patents in Italy, Germany, the United States, the United Kingdom, and Japan. When Sony started mass-producing its Walkman in 1979, Pavel sought well-deserved royalty fees, which he eventually got in an out-of-court settlement in 2003. Above all, he seems to grasp the transformative importance of his invention. "I was in the woods in St. Moritz, in the mountains, the snow was falling down, I pressed the button and suddenly we were floating," he recalls of the first time he used it. "It was an incredible feeling, to realize that I now had the means to multiply the aesthetic potential of any situation."[41] William Gibson adds, "I haven't had that immediate a reaction to a piece of technology before or since. I didn't analyze it at the time, but in retrospect, I recognized the revolutionary intimacy of the interface. For the first time I was able to move my nervous system through a landscape with my choice of soundtrack."[42] After perhaps the book, the Walkman was both the next truly personal medium and the next step in our creation of a virtual reality. "It's almost a metaphorical realization of what we do anyway—a souped-up version of being lost in one's own thoughts. Given the conditions in which most of us increasingly live and the hard work required to obtain personal space, I think that's not entirely a bad thing."[43] Meaning-making happens at the interface between technology and culture.[44] As sound studies scholar Michael Bull writes, "Almost any experience can be construed as filmic by personal-stereo users."[45]

To be sure, there were dissenters: "I don't think it's a good thing to make your life like a movie," one of Bull's interview subjects observed.[46] "It emphasizes the step of removal from where you are."[47] The Walkman gave us access to sounds from the past regardless of physical location, allowing sound to shake off its original context for the listeners' own purposes.[48]

It's nearly impossible to talk about a technology such as the Walkman without mentioning its relationship to memory and nostalgia—not only the nostalgia of previous eras of technology but also the nostalgia inherent in the very practice of recording and replaying pieces of the past. Nostalgia implies false or "imagined" memories, memories that are devoid of significance that we fill with what we imagine they were like.[49] Paul Grainge points out an important distinction between nostalgia as a commercial mode and nostalgia as a social or collective mood. The former is often enabled by the latter as we drool over reissues of long-lost demo tapes or clamor for reunion tour tickets.[50] Thanks to recording technology, we live in an era when, as Andreas Huyssen puts it, "the past has become part of the present in ways simply unimaginable in earlier centuries."[51] The nostalgia implicit in the Walkman is the result of its ability to remove and appropriate the historical context of recorded music. It can take you back to where you were when you first heard that song. In this way, the Walkman is, as Iain Chambers describes it, "simultaneously a technical instrument and a cultural activity."[52] The French psychoanalyst and political philosopher Félix Guattari put it this way: "Human beings have been fundamentally deterritorialized. Their original existential territories—bodies, domestic spaces, clans, cults—are no longer secured by a fixed ground; but henceforth they are indexed to a world of precarious representations and in perpetual motion. Young people are walking around the streets with *Walkmans* glued to their ears, and are habituated by refrains produced far, very far, from their homelands."[53]

The Walkman's cultural significance also owes something to its similarities and differences from other mobile devices.[54] It is similar to the iPod in that both devices provide portable access to a private experience: listening to music on headphones. It is different in the affordances it presents to facilitate access to that experience. The Walkman is similar to the boombox in that it is a portable source of sound. It is different in the way that it arranges the people and the sound in space. As Joshua Meyrowitz notes, "There is a big difference between listening to a cassette tape while driving in a car and listening to a radio station, in that the cassette tape cuts you off from the outside world, while the radio ties you into it."[55] It is a shift not so much from public to private but from public to personal.

"Flashback to Eno, 1980," writes the producer Daniel Lanois in his 2010 memoir, *Soul Mining*. "The introduction of the Walkman was the beginning of the 'on the go' personal entertainment station. I can remember a dinner-table Eno rant. The Walkman promised mobility but introduced isolation."[56] That's the engineering tradeoff of personal media. As every new contrivance frees us from something, it binds us to itself. *Here, with this gadget, you can connect to people and information from anywhere, but you can never leave the gadget behind.* And now, as Lanois continues, "every face is buried in a computer, every ear is plugged with headphones."[57] Again, our devices are divisive.

When I got my first Walkman and stopped lugging around my boombox, it was a blessing not only to my back but also to the sanity of those around me, most notably my parents. Wherever you were, the boombox created a sonic space. No place to dance and hang out with friends? Pop in the mixtape and blast the box: sounds for all. That is unlike the iPod, which according to Bull, "appears to have furthered the culture of mediated isolation whereby connectivity is increasingly engaged with absent others," not to mention smartphones and streaming services.[58] The presence of the boombox was a public presence and not necessarily a welcome one. Mack Hagood points out, "Like hip-hop, culturally diverse ways of being in the world are encouraged when they can be commodified and consumed as media, but are rejected as disruptions to the smooth circulation of capital when practiced in lived spaces such as airports, airplanes, or city streets."[59] The cassette-enabled boombox brought private listening preferences right out into the public.

In spite of the power of the boombox to create an impromptu space of sound, the cassette tape, which was cued up to challenge the LP and the 8-track market as the musical format of choice, didn't take off until it became hypermobile thanks to the personal portability of the Walkman. High-quality sound once required equipment too cumbersome to be carried. The cassette in concert with the Walkman gave everyone their own soundtrack. Their portability and customizability made the Walkman *the* personal item for music fans and

1979 advertisement for the Sony Walkman TPS-L2.

their friends. On road trips, tapes were as much a part of the journey as the road.[60] The portability and recordability of cassettes, of which all sounds so very labor-intensive now, made them the precursor to today's MP3s and streaming services. The individual soundtrack provided by personal media undermines the narrative of the city, allowing us each to form our own story, our own "re-combinant city" in Gibson's phrase.[61] Just as the book individualized the exchange of stories and information, the cassette tape and its attendant technologies individualized music listening.

When the Walkman first came out, it was intended for sharing. The first models released had two headphone jacks. I distinctly remember the first one I listened to having dual jacks. Nevertheless, initial sales numbers indicated that few users were sharing their devices, so Sony switched its approach. In the ads, Weheliye writes, "Couples riding tandem bicycles and sharing one Walkman were replaced by images of isolated figures ensnared in their private world of sound."[62] This source of secret sound, the black box of the personal stereo, is where its cultural meaning truly lies.[63] "People who walk around with a Walkman might seem to signify a void, the emptiness of metropolitan life," writes Iain Chambers, "but that little black object can also be understood as a pregnant zero, as the link in an urban strategy, a semiotic shifter, the crucial digit in a particular organization of sense."[64] Sony ceased production of the cassette edition of the Walkman on October 25, 2010, but the internet, which has been feeding MP3s to portable players since the late 1990s and now streams music directly to smartphones, continues its legacy. Whereas the television brought mass-produced narratives into our home, conflating public and private in its own way, the Walkman allowed us to create our own private narratives in public.

FANTASTIC DAMAGE

There's a perverse comfort in broken machinery.
—Ellen Ullman, *Close to the Machine*

Creating his own narratives with media, Christian Marclay has made a career of repurposing records. The composer and artist John Cage, who despised the idea of recorded music, once said, "Music instructs us . . . that the uses of things, if they are meaningful, are creative; therefore, the only lively thing that will happen with a record, is, if somehow you could use it to make something which it isn't."[65] Taking Cage to heart, Marclay's extensive body of work includes rearranging record covers into composite, Exquisite Corpse–like images; lining gallery floors with records for gallerygoers to walk on and then playing those trampled records later; and, of course, manipulating records on turntables in a Duchampian take on hip-hop turntablism to create new sound collages. "I've always been interested in all of the sounds that weren't recorded on the record, but were a result of mishandling, of time, of damage," he says.[66] Having come up in the same New York scene as improvisors like John Zorn, artists like Laurie Anderson and Dan Graham, and no wave bands like James Chance and the

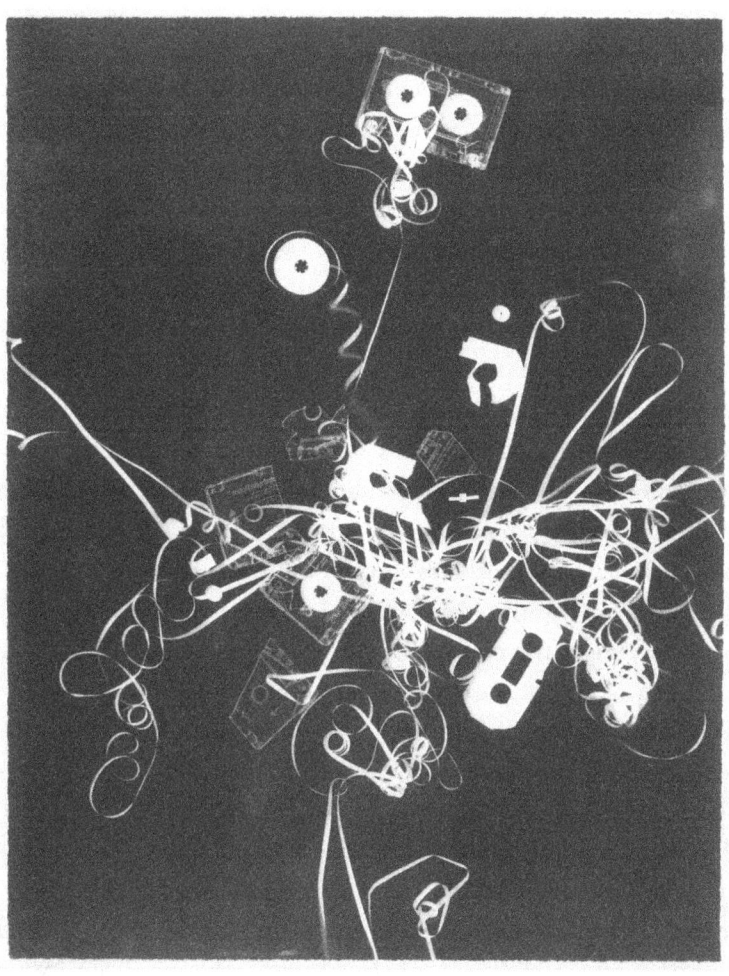

Christian Marclay, Untitled (R.E.M. and Sonic Youth), 2008. Cyanotype on 156lb. Cold Press Aquarelle Arches, 26 x 21 in. (66 x 53.3 cm). © Christian Marclay. Courtesy of Paula Cooper Gallery.

Contortions, DNA, and Mars, Marclay cites punk as the impetus for his sound explorations.

"The punk movement was a liberating influence," Marclay says, "with its energy, its non-conformism, its very loud volume of sound. Its amateurish, improvised side gave me the courage to make music without ever having studied it."[67] Marclay is more artist than musician. Where hip-hop and club DJs work with carefully curated crates of vintage vinyl, Marclay finds old LPs in junk shops and broken records in the street, the detritus of soundtracked days long past. He tells Kim Gordon, "I went to art school, not to music school. I don't think like a musician."[68] He's worked in many other media besides sound. Readymades, collages, video, and performances all find their way into his work. "The more I worked with records," Marclay says, "the more I realized the potential of all the sounds generated with just a turntable and a record and started to appreciate all these unwanted sounds that were traditionally rejected: skipping, clicks and pops, all this stuff that people didn't want. I started using these sounds for their musical quality and doing all kinds of aggressive, destructive stuff to the records for the purpose of creating new music."[69] Like most of us, Marclay came up primarily listening to recorded music, with live performance serving a different purpose. Having no real musical skills, he saw records and turntables as a way to do live performances. That is, instead of passively listening to recordings, he wanted to put records in a more active role. Vinyl records are fragile, though, so that became a part of the work.

"I try to make people aware of these imperfections, and accept them as music," he says. "The recording is a sort of illusion; the scratch on the record is more real."[70] He questions each medium in terms of itself, not only revealing its flaws but focusing on them.[71] Marclay uses the broken metaphor, "Memory is our own recording device," but his works can be seen as montages of such metaphors.[72] There seems to be something damaged or at least slightly off about the connections he makes and breaks. The damage and discontinuity inherent in these media—records, tapes, telephones—are the raw materials that Marclay works with.

"The loss of control in music is actually what interests me the most," Marclay says.[73] The back and forth between control and loss thereof is the backbone of improvised music. Greil Marcus describes punk as remaining "suspended in time," an unfinished nihilistic revolution.[74] And just as punk has a dubious re-

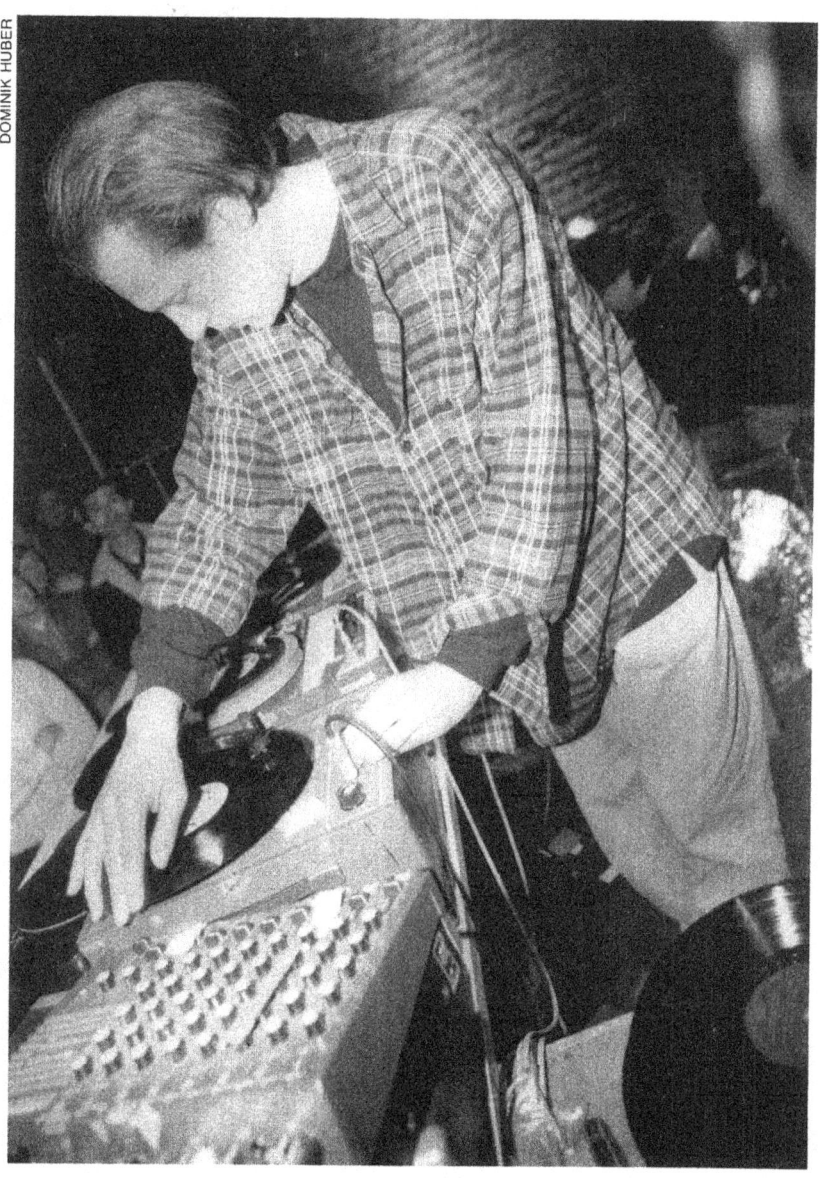

Christian Marclay making a mess of the medium.
Photo by Dominik Huber. Promotional photograph.

lationship with Dada and the Situationists, so does Marclay. The Situationist theorist Guy Debord's first book, 1959's *Mémoires*, was bound in sandpaper in order to wreck the books shelved next to it. The noise-punk band White's 1996 seven-inch single "Life on the Ranch of Elizabeth Clare Prophet" was issued in a self-destructive sandpaper sleeve. In 2002 the composer William Basinski released *The Disintegration Loops*, a four-album set composed of lengthy tape loops that revealed more and more of their flaws as they played. Christian Marclay released *Record Without a Cover* on his own label Recycled Records in 1985. One side was playable, encoded with recordings of manipulated records; the other side was pressed with text of the titles and the instruction "Do not store in a protective package."[75] It was just what its name implies, with a similar thought in mind: exposed to the elements and wear of everyday life, the record sounds different every time you play it. Its slow decay will become a part of its performance. Leaving the record without a sleeve and thus unprotected was an act of creative destruction, artifacting in action.

Though they were touted as a solution to those flaws, digital formats like CDs typically stop functioning when broken, scratched, or otherwise damaged, their digitized works trapped inside or lost to the ether. Analog artifacts endure.[76] Paul Hegarty writes, "For Marclay, the record is not just being deviated from its original use; it comes from disuse (being discarded) and inutility (damage), and it will continue into something formless. If such works are then put onto CD, the process might be over on a literal level, but the sounds will be a constant parody of the perfection attributed to CDs, a reminder of the hopelessness of attaining perfection."[77]

By using analog records, Marclay is able to pull the listener out of the recording by foregrounding the flaws of the medium.[78] By flaunting their flaws, analog media remind us of our own connections to the physical world.

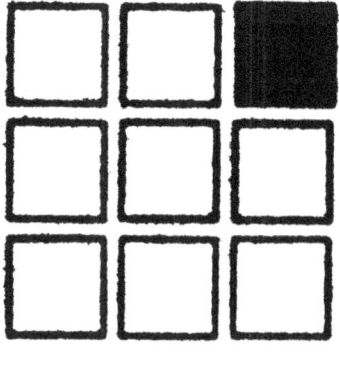

Algorithm Nation

Maybe a record was always a distorter of memory, not to be
sought. Better to hear by way of reenactment.

—from Kim Stanley Robinson's *2312*

I know you all think you live in all the times at once, everything
recorded for you, it's all there to play back. Digital. That's all that
is, though: playback. You still don't remember what it felt like.

—Shinya Yamazaki, in William Gibson's *Virtual Light*

Culture is good at pointing to things and calling their
name but not so good at describing the relationships
between things or the repertoires they enact.

—Keller Easterling, *Medium Design*

CIRCULAR WINGS

People used to believe in circles. They came to believe in lines.

—Douglas Rushkoff, *Team Human*

The first time I heard a compact disc was in middle school. My best friend's dad had just replaced his entire collection of LPs with CDs. They sat in stacks beside the apparatus that played them. They were like extraterrestrial objects, something from the science fiction we were into at the time. They were also off limits. We were not allowed to touch them.[1]

One day in 1983 my friend's dad sat me down on the couch in the middle of their den. The angled sunlight of autumn streaked through the limbs and leaves of the trees in their small front yard. Four large brown cabinet speakers, sitting one each in the corners of the room, were all pointing directly at me. He put on "Owner of a Lonely Heart" from Yes's then-new record, *90125*, loud enough for all the neighbors to hear. The opening samples stumbled around the room before the lead guitar took hold. That first horn stab, a sample from "Kool Is Back" by Funk, Inc., which is itself a cover of "Kool's Back Again" by Kool and The Gang, a leftover from Trevor Horn's *Duck Rock* sessions with Malcolm McLaren, sounded like a laser shot from space. I remember being able to feel Chris Squire's bass thumping through the hardwood floor as Trevor Rabin's guitar swirled around the samples that bounced off the walls and my skull to dizzying effect. That day the CD earned and maintained its otherworldly reputation in the history of recording formats, supplanting the raggedy cassette and the outmoded vinyl record.

VICARIOUS LIFE

The fault is not in our tools, we say, but in ourselves.

—John Gray, *Straw Dogs*

Prior to the phonograph, listening to music was strictly a social practice. People gathered together to hear others play instruments and perform.[2] Even playing an instrument by oneself was frowned upon.[3] "We tend to think of music as this sightless thing because recordings have distorted our understanding of music," Christian Marclay says. "We think it's this totally abstract thing that comes from nowhere and just exists as a recording, a disembodied invisible performance."[4] We're so versed in the lexicon of media that we rarely stop to think about how or why it's made. By the time Thomas Edison patented the phonograph in 1878, the public was not only familiar but had become comfortable with the idea of preserved foods.[5] Thus, "canned music," as the composer John Philip Sousa put it, was ready for mass consumption.[6] Picturing a world without such prepackaged media is difficult to do now. We preserve everything ourselves, or we expect someone to preserve it. The problem is not so much the authenticity of our entertainment and information but more how to parse the sheer expanse of it. Once we're able to record something, we start to feel the risk of missing out on capturing it.[7]

In the first chapter of his 1992 book *Turn Signals Are the Facial Expressions of Automobiles*, the design critic Donald Norman described going to see a sixth-grade play in a relatively small auditorium. "If there had been only fifty parents present, it would have been crowded," he wrote. "But in addition to the parents, we had the video cameras."[8] This anecdote describes a time well before the camera shrank and merged with the mobile phone. Video cameras were cumbersome, and many didn't yet run on batteries and needed wall outlets to function, hence his concerns about space. Norman continued, "Ah yes, once upon a time there was an age in which people went to enjoy themselves, unencumbered by technology, with the memory of the event retained within their own heads. Today [ca. 1992] we use our artifacts to record the event, and the act of recording becomes the event."[9]

Norman called this need to record instead of paying attention "vicarious experiencing." As wireless networks and their bandwidth got broader and smartphones and their cameras got smaller, sharing experiences got easier and easier. Echoing Norman many years later, the comedian Doug Stanhope used to do a joke about people recording his stand-up shows. Having always not only allowed but encouraged the sharing of his material online, Stanhope grew tired of people filming his performances, taking quotations out of context, and posting incomplete bits. "Hey, why don't you just watch the show?" he asked. His mocking response: "Nah, I'd rather watch it on my laptop later."

Similarly, on the way to record their 2006 record, *Blood Mountain*, with the Seattle producer Matt Bayles, Atlanta metal band Mastodon test-drove many of their new songs in front of live audiences across the country. It was the last opportunity they would have to do so. By the time they were ready to record their next record, 2009's *Crack the Skye*, camera-enabled smartphones and easily uploaded video made it impossible to practice in public. No more sampling a song in front of a live audience, lest a lesser version be shared with the world before it's ready.

You know how it feels when your boss or a friend or colleague asks to see something you're working on before it's finished. You have to make excuses about every angle you were trying and every flaw you already know needs fixing. The unfinished can be a fragile state for the creative process, and the unfinished work is only shades of its future self at best. The unfinished opens the creator to questions that the finished version would never elicit. Consider the same situation for comedians like Stanhope. Jokes have to be worked out in front of audiences. The only way to truly test the wording, the phrasing, the cadence of a set-up, and the optimal delivery of a punchline is to hear a live crowd react. There's no way to do that if you run the risk of having every joke-in-progress recorded and posted online. And having an online audience hear all of your rough drafts diminishes the final version's impact. Everyone loses.

WAR@33.3

Music, like drugs, is intuition, a path to knowledge. A path? No—a battlefield.
—Jacques Attali, *Noise: The Political Economy of Music*

With audio recording, Edison had business communication in mind. He saw the phonograph as a more efficient, more accurate replacement for the practice of stenography, which entailed human labor. But by the beginning of the twentieth century, phonographs were being marketed as musical playback devices.[10] Whereas live musicians had once accompanied silent films in theaters, by the 1930s canned music was characterized by anxious musicians as a human-replacing, robotic monster. The American Federation of Musicians formed the Music Defense League and ran ads decrying recorded music as a mechanical menace. In 1931 their president, Joseph N. Weber stated: "The time is coming fast when the only living thing around a motion picture house will be the person who sells your ticket. Everything else will be mechanical. Canned drama, canned music, canned vaudeville. We think the public will tire of mechanical music and will want the real thing. We are not against scientific development of any kind, but it must not come at the expense of art. We are not opposing industrial progress. We are not even opposing mechanical music except where it is used as a profiteering instrument for artistic

A robot grinds musical instruments to make canned music. *Syracuse Herald*, 1930.

MAKING MUSICAL MINCE MEAT!

BIFF, bang, crash! . . . The wheels turn . . . the cogs mesh . . . Canned Music fills the air!

debasement."[11] Of course, now there often isn't even a person who sells you your ticket.

Creators of culture often lose control of their art when certain technologies step in. Some adapt, some adopt, and others disappear. The first records appeared in the early twentieth century. By the 1940s, a standard was set, and an organizing principle was established. The standard was thirty-three and one-third revolutions per minute, with about forty-five minutes of recording time, approximately twenty-three minutes per side. Traditionally, most artists recorded ten or so three-to-four-minute songs, which fit nicely into the LP format. But artists who bristled at the restrictions of the popular song could record single pieces that lasted the entire side of a record—and they did. Think the lengthy improvisations of John Coltrane or Miles Davis or the proggy self-indulgence of Yes or Rush.

In the 1970s, punk rock threatened not only the dinosaurs of the dark side of the moon but also the format itself. Punk's working-class angst had no time for the extravagance of prog and festival rock, much less with the length of their songs. The spiky bursts of punk's aggression led to a rise in significance of the seven-inch single. Bands were heard today and gone tomorrow, rarely together long enough to record an entire LP's worth of material, and fans rarely expected much more.[12] Their pithy musical statements didn't require such a long run time. Theirs was an extremist gesture—to appropriate Simon Reynolds, "the kind of cultural spasm that can only exhaust itself."[13] But the LP format and its limitations on recording, production, and listening remained in force for the next fifty years.

FAST-FORWARD

In conditions of digital recall, loss is itself lost.
—Mark Fisher, *Ghosts of My Life*

In 1988 when the Cure was recording its eighth record, *Disintegration*, band leader Robert Smith said it was the first time that they had gone into the studio knowing that they would be recording for a release on compact disc, which meant they could plan to record over an hour of music. "*Disintegration* is the first real CD-LP," he claimed. "It was about time the musicians learned to use this format: instead of two twenty-minute sides of an LP, you now have a seventy-minute stream of music without interruptions."[14] The LP had restricted bands to a runtime of forty-five minutes, but with the advent of the CD, artists had additional time, and they could include "bonus tracks" on CD reissues of older LPs.[15]

Just as importantly, the CD changed the act of listening. Notions like "fast-forward" and "rewind" lost their meaning overnight. Digital media changed listeners' access to the individual songs on a record, and it changed their expectations as well. "The history of mobile listening," writes Michael Bull, "is also the history of ratcheting up of consumer desire and expectation—consumers habitually expect these technologies to do more and more for them."[16] The CD was just the next physical format for the commercial distribution of music. Its aesthetic proximity to the LP and the likelihood that it is music's last physical, commercial format have caused some to conjecture that it will end up like the LP, as a cultural artifact to be revered, preserved, collected. "The tangibility of the CD is part of its charm," writes Mark Katz. "A collection is meant to be displayed, and has a visual impact that confers a degree of expertise on its owner."[17]

Without physical formats, how do you let someone know what you like? How do *you* even know what you like?

Dischord Records founder Ian MacKaye expressed an alternative view to me in 2009: "I don't think the 'decline of the compact disc' has affected the way I approach the idea of recording music. We just press fewer CDs, more vinyl, and make it available online."[18] In 2006 CD sales were still strong, generating $9.4 billion. That was down to $3.4 billion by 2010, and soon the industry dropped to its lowest point since 1990.[19] To put it another way, sales of compact discs dropped by half between their peak in 2000 and 2007, and by 97 percent in the years since. By 2020, the CD made up only 4 percent of total music industry sales.[20] "If anything," MacKaye continues, "I reckon the fact that so many people are listening to music on the earbuds and headphones that come with personal listening devices has made me think more about the recording process. I don't listen to music that way, and I wonder if my aesthetic still 'works' when shot directly into the ears. Having said that, I don't have any intention of doing anything differently."[21]

"On a conscious level, the decline of the compact disc has had no effect on how I record music," K Records founder Calvin Johnson similarly told me in 2009. "On the subconscious level, it has filled me with glee."[22] Like Johnson, others see its small size and subsequently small cover art, as well as its mass-produced disposability, as signs it will go the way of the cassette and the 8-track, detritus left behind by technology's forward momentum, moving at the speed of the instantly downloadable MP3. The pro skateboarder and recording artist Duane Pitre adds,

> I like to have something in my hands. I am not very fond of MP3s as a final product as sound quality is compromised, and you have to interact with a computer. Also, there is no artwork to gaze at, no booklet to flip through, etc. Music used to be about an "escape" of sorts for me (still is sometimes) and being locked onto computer is by no means an escape in my eyes. But anyhow back to the ways to release . . . I am very fond of vinyl releases with free MP3 download coupons. I think that is the best of both worlds.[23]

"MP3" stands for MPEG Audio Layer III, after the encoding process that was under development by the Moving Picture Experts Group (MPEG) for ten years before it was brought to fruition in 1992. Identifying the internet as its native environment, the "dot-mp3" file extension was born on July 14, 1995. "At some point in the late '90s," says Karlheinz Brandenburg, whose PhD work in 1982 landed him in the middle of the development of the format, "MP3 was technically the best system out there and at the same time it was accessible to everybody."[24] These two factors gave the MP3 an early advantage, and every device that plays digital audio files was subsequently designed for the format. Once online file sharing and the iPod came online around the turn of the millennium, the floodgates were opened, and music was liberated not only from the dams of physical formats but also physical spaces. What once took rooms of equipment and stacks of physical media to enjoy is now in everyone's pocket. With the introduction of the first portable MP3-player in 1998, the Recording Industry Association of America (RIAA) started its legal battle against the digital revolution.[25]

While legitimizing the downloading of music was one of the hurdles that Apple faced with record labels when introducing iTunes in 2001, another was the breaking up of the then-standard CD into individual songs. Record labels had long been making billions from record, tape, and CD sales. Individual hit songs were mere marketing, advertisements for the main product. What Apple wanted to do legitimized downloading but broke down the price structure of the millennial record industry. Instead of buying fifteen to eighteen songs for $17.99, now listeners could buy just the song they wanted for 99 cents. It streamlined the process for listeners but cut out massive profits for the labels.

ALGORITHM NATION

As the infrastructure has shifted, so has the sound. Songs made for streaming are shorter, some say due to diminishing attention spans. Others say it is to game the streaming systems themselves. The quantifiable aspect of music consumption is now the number of plays a song gets. A shorter song presumably leads to more plays and is more likely to be spread via the algorithms on streaming services. Singles in the radio era were traditionally three to four minutes. Streaming internet singles hover around two and a half.[26] The limitations of platforms play a part too.[27]

The first record by Philadelphia artist Tierra Whack, *Whack World*, is a fifteen-minute audiovisual suite with fifteen one-minute compositions. Those limitations might sound arbitrary, but, when Whack released the project in 2018, the image- and video-sharing social media app Instagram had a one-minute cap on video posts. Whack made this project for release specifically on Instagram. Claiming she wanted to give a glimpse of her many creative and emotional sides, Whack also acknowledges the attention-deficit issue. She tells the *New York Times*, "My age, my generation, we get bored so easily. I know how I am—I'll listen to a new song, and I only want to hear 30 seconds of it before I tell you, 'Nope—trash.' I have a really short attention span, but I have so much to offer. I wanted to put all of these ideas into one universe, one world. I'm giving you a trip through my mind."[28] Like the shortest of episodes in the shortest of seasons, the songs and videos of *Whack World* stand individually even as they add up to a whole. That these formats are becoming more and more ephemeral also seems to be a part of their evolution.

A few years ago I was having lunch at a bar in Chicago when an Archers of Loaf song came on over the speakers. Excited, I told my partner that I was a big fan, described the first time I saw them (at the Crocodile Café in Seattle), and said that I had seen them a dozen or so times during their first run in the 1990s, once even traveling to Vancouver to see them play with Treepeople and Spoon. I told her how, fancying myself an indie-rock mogul, I had plans to put together a compilation of Chapel Hill bands, and they were the first to agree to contribute. And how I'd gotten to be pretty good friends with their bass player, Matt Gentling, how he's also a rock climber, and how we had stayed in touch over the years.

As we continued to eat, song after song of theirs came on. Now, as great as they are, the Archers are not a normal thing to hear in public, much less several of their songs in a row. I went up to the bar and asked who the Archers of Loaf fan was. No one knew what I was talking about. I explained the same thing I just explained to you, that hearing that many of their songs in a row wasn't normal. The third person I asked said it was just a streaming service. Somehow the streaming service's algorithm had gotten stuck on the Archer of Loaf catalog. Not a bad place to get stuck, as far as me and my lunch were concerned, but it irked me that I was alone in the experience I was having.

Recommendation algorithms are opaque and proprietary pieces of code. Each streaming service has its own and guards it aggressively. The algorithms themselves are nested black boxes. There's no way to know exactly how they

Matt Gentling of Archers of Loaf on the cover of *Fizz* in 1994.
(If you look closely, you can see a *Front Wheel Drive* sticker
on his bass. That was my zine at the time.)

arrive at their results, but most of them get there via a few basic metrics. Each service collects your listening habits, builds a profile of you as a listener, then recommends artists, albums, and songs based on your profile. There are also genre tags and other metadata, filters based on similar listeners, and platform-generated playlists or channels, but there's no way to know what criteria pushed the next song out of the database and into your ears.

Music fans of a certain age extol the effort it took in the pre-internet era not only to find the good stuff but also to learn about it in the first place. We relied heavily on curated spaces that are less and less influential now where they exist at all. Local record stores, underground press, small labels, narrowly focused radio shows, and genre-specific venues created community and shared knowledge. The sound that 1990s Seattle is known for seems to be the last historical example that emerged from such a community, formed unfettered or influenced by outside forces. For example, the movement's flagship label, Sub Pop, started as a zine, a compilation cassette series, and a column in *The Rocket*, one of Seattle's local music papers. It was a special time and a special place. Not only did all of this happen in a time untainted by the internet, but the Pacific Northwest is also isolated geographically from the paths of nationally touring bands. I moved to Seattle in the summer of 1993, after most of the world knew about what was going on in the area but before it peaked.

I do not lament the effort it took to find new music then, but the missed connections like my Archers of Loaf lunch don't happen when you're truly sharing an experience. It's one thing to be cut off from each other by choice. It's truly another when we don't even choose it.

WE CAN FORGET IT FOR YOU WHOLESALE

You might have mass storage containers—cassettes, VHS tapes, floppy discs, and so on—that lack a device capable of reading them, ghosts of information past trapped in a black box forever.[29] We're all amateur archivists whether we notice or admit it, but access to our archives has an expiration date. If you tour the back rooms of archives such as the Harry Ransom Center on the University of Texas campus in Austin, you'll see stacks and shelves full of canisters of unwatchable film. Some of it is missing a proper apparatus with which to be viewed, and some of it is slowly deteriorating, melting in its own fumes.[30]

Off-loaded, outmoded, entire histories hidden in halted formats, recorded yet irretrievable, remembered by media yet forgotten by technology.

In 1981 computer pioneer Ted Nelson posed the main problem of the archive: impermanence. "Books disappear. Knowledge of the past is lost. Libraries *burn*, and each time, we are diminished," he writes in *Literary Machines*.[31] The renowned studio engineer Steve Albini was known for adhering to analog recording practices in an industry that has, for better and worse, largely shifted to digital techniques. Audiophiles will tell you that analog recordings sound better, but Albini's reasons for staying analog were archival: "People's life work—their sweat and effort, their entire creative soul—is being recorded on something which will not be playable in a few years' time. It's essentially the same as cutting somebody's record into ice."[32] A digital format doesn't exist that will maintain a permanent document of a recording session. Digital formats degrade, become unreadable, or are proprietary, leaving behind "orphaned masters" no one can access. Albini's devotion to analog methods stemmed from his interest in providing musicians what they'd paid for: a permanent recording of their work.

"A key form of data is *music*," Nelson continues. "The overall problems of musical archiving, cataloging, annotation and scholarship have not had a unified software base from which they may be correlated with performance."[33] Nelson was proposing a system not only to maintain such recordings but also to annotate and keep up with cover versions, samples, and other derivatives. Like Albini, Nelson is interested in protecting the original work. "The recording part is the part that matters to me—that I'm making a document that records a piece of our culture, the life's work of the musicians that are hiring me," Albini said. "I take that part very seriously. I want the music to outlive all of us."[34]

I struggle to get my students to take notes. I require notebooks in all of my classes, and I regularly reiterate the benefits of handwritten notes. Some of them catch on, but most rely on other means. I get it. Theirs is an on-demand culture. Their world is one of on-demand entertainment, goods, services, and information. Why should they bother to write anything down? If it's not something they need right at that moment, anything is available at any time. An archive of everything ever is right at their fingertips, so they think. Nelson continues, "But who is to control and safeguard these systems becomes the next question."[35] When we rely on others to keep up with our files, we place a lot of trust in people who don't prioritize what we find precious.

In June 2008, a fire in Building 6197 at Universal Studios wiped out the original master tapes for over half a million songs, including hits and favorites by Lefty Frizzell, Loretta Lynn, George Jones, Merle Haggard, Quincy Jones, Neil Diamond, Sonny and Cher, Captain Beefheart, Iggy Pop, the Police, George Strait, Steve Earle, Elton John, John Coltrane, Common, Oingo Boingo, Eric B. and Rakim, Snoop Dogg, Tupac Shakur, Eminem, Nirvana, R.E.M., Sonic Youth, and Nine Inch Nails, among many, many others.[36] This was an undeniable tragedy, the abject failure of an archive.

A decade later, erstwhile social-networking site MySpace lost twelve years of its users' music. "As a result of a server migration project, any photos, videos, and audio files you uploaded more than three years ago may no longer be available on or from MySpace," they said in a statement.[37] "We apologize for the inconvenience and suggest that you retain your back up copies." The lost files include everything uploaded between 2003 and 2015, over fifty million songs by fourteen million artists, as well as countless photos and videos. "Due to a server migration," the company informed thousands of horrified musicians and fans, "files were corrupted and unable to be transferred over to our updated site. There is no way to recover the lost data."[38] If it sounds sloppy, it probably was.

"I am deeply skeptical this was an accident," says Kickstarter CTO Andy Baio. "Flagrant incompetence may be bad PR, but it still sounds better than 'we can't be bothered with the effort and cost of migrating and hosting 50 million old MP3s.'"[39] Ted Nelson goes on to point out the lessons of Ray Bradbury's 1953 novel *Fahrenheit 451*, in which a group of people, under an oppressive regime where literature is banned, take on the task of memorizing books before they are lost. "This allegory may take on new meanings in a world whose future rulers would like to alter the past—a common conceit in such occupations."[40] It's easier to revise history if there's no record of it in the first place.[41]

Though neither of them was the burning of the Library of Alexandria, the Universal fire and MySpace server migration touch on a few of the implications and the importance of maintaining the archive. The selection of particular artifacts or information to be saved or archived is an act that predisposes archivists and archaeologists of the future, not to mention everyday consumers, to give it special consideration in years to come.[42] Then too, what we record receives future attention just by dint of being recorded. We think of archives as collections of pieces of the past, but we use them to save those things for future use and retrieval. Moreover, meaningful archives take a long time to amass and no time to destroy.[43] We've never really had a problem with storage space. What we have is a persistent problem with the preservation of what we store.[44] That's why it matters who stewards not only the information to be saved but the technology that records and archives it as well.

The health of our archives also has cultural impacts. The anthropologist Greg Urban writes, "The movement of culture through the world is possible only because it becomes lodged, however fleetingly or enduringly, in material, perceptible things—stainless steel pots, girdled bodies, and even words as physical objects, as sounds or shapes. These are things that humans have fashioned, and hence that make tangible or manifest accumulated social learning. It is from such physical manifestations, and only from such physical manifestations, that culture is able to journey through the world, making its way from individual to individual, group to group."[45]

When groups of people rely on the stewardship of these artifacts by outside forces, their own culture suffers. Urban writes regarding tribes that once made their own ceramic pots and came to rely on metal pots made by others.[46] The reliance on outside technology weakens the internal culture. Our personal media need to give us access not only to cultural artifacts but also to our personal archives.[47] We cannot trust technology companies to handle our data. Even with all of our precious gadgets intact, we need to rely on ourselves more and others much less. The impacts of personal archival practices are evident when implemented properly.

BREADCRUMB TRAIL

Emerging from Washington, D.C., in the late 1980s, the band Fugazi maintained a vehemently independent career working according to their own set of hard-won principles, working in the music industry only when they had to, and adopting and adapting the changing technologies as needed. Former Minor Threat front-person Ian MacKaye had been in a few bands in the meantime—Embrace, Egghunt, Pailhead—releasing a smattering of singles, an LP, and an EP. So, when Fugazi's first self-titled EP came out, no one really expected it to be the beginning of the biggest thing Dischord Records would produce. Fugazi went on to be the very definition of the independent underground in the 1990s.

The author's Minor Threat and Fugazi cassettes.

Minor Threat was one of my favorite bands of the early 1980s. I had two of their records on old black cassettes with paper labels. I wore them out by the time the first Fugazi EP came out in 1988. Often cited as an offshoot or sub-genre of seventies punk, American hardcore blazed a new trail for underground aggression. The first song I remember hearing described as "hardcore" was "Richard Hung Himself" by D.I. The stripped-down, stark anger and energy of that song stuck with me. Its clipped lines and shouted vocals seemingly spoke directly to *me*. Not long after, I found Bad Brains, 7 Seconds, Youth Brigade, and Minor Threat, among others. From the moment I heard it, the raw sound of hardcore became a touchstone for my listening.

Along with the sound, the look was different too. Colored Mohawks, spiked leather jackets, and combat boots were replaced with buzzcuts, T-shirts, torn jeans, and Vans or Chuck Taylors. Skateboarding and music melded. *Thrasher* magazine started releasing their *Skate Rock* compilations of underground hardcore bands. No need to fix your hair or lace up your boots. The dress code for skateboarding and hardcore was the same, so you could skate from the session directly to the show.

While MacKaye was busy with Minor Threat, Guy Picciotto and Brendan

Ian MacKaye on stage with Fugazi at the Sylvan Theater in the shadow of the Washington Monument, August 7, 1993. Photo by Pat Graham. © Pat Graham Photography. Courtesy of Pat Graham.

Canty were in a band called Rites of Spring, but by the mid-1980s both groups had run their courses. Playing guitar and bass respectively, MacKaye and Joe Lally started Fugazi as a three-piece with Canty on drums. A fan of the band and a fixture at their early shows, Picciotto was the last to join. They played their first show on September 3, 1987.[48]

Where Minor Threat helped define what became hardcore and Rites of Spring was one of the first emo bands, Fugazi was past all of that from the beginning. The bedrock rumble of Lally and Canty gave the dual guitars and vocals of MacKaye and Picciotto room to roam. They were post-punk in the arty and aggressive spirit of its original form. As likely to lull you with melody or near silence as they were to pummel you with guitar or rhythm, Fugazi was as fresh as they were familiar. Post-punk has always been more lenient than its punk precursor. It's a genre more forgiving of experimentation, more tolerant of outside influences. Where, as writer Andy Greenwald points out, punk is a cul-de-sac, post-punk is more of a wide, winding road.[49] At its commercial peak in 1979, it included the angular and atonal guitars of Gang of Four, Public Image Limited, and Wire, the damaged pop of Joy Division, Talking Heads, and Devo, and the gothic drama of the Cure, Sisters of Mercy, and Siouxsie and the Banshees. Like all genres, there's an invisible thread that loosely unites these bands, an unseen aesthetic that links them. That thread stretches from post-punk's origins in the early 1970s through Fugazi and beyond.

Some of the principles by which Fugazi conducted itself included five-dollar tickets, no set lists, no moshing, and no merch. Just as they limited their involvement with the music business, they also limited their gear. "I'm not interested in options," MacKaye says, echoing Brian Eno's concerns about constraints noted earlier. "I'm interested in how far I can take this simple equation, which is an amp, a cord, and a guitar, and how much I can do with it."[50] Fugazi went on tours without a record available, working songs out on the road, and winning fans across the United States and in Europe one punk house or church basement show at a time.

They released their records exclusively through Jeff Nelson and MacKaye's Dischord Records where they set their own prices. Fugazi didn't do interviews with publications they didn't read, and Dischord still only advertises with the smallest of zines and magazines. Outside of the music on physical formats, they offered no extraneous merchandise, which included no T-shirts. The most popular bootleg Fugazi T-shirts even read "This Is Not a Fugazi T-Shirt." Prefiguring the generation-defining mantra of *Fight Club*, on their song "Merchandise" they sing "You are not what you own."[51] Fugazi emerged as physical formats were still evolving, and the compact disc had yet to take over completely. Their first two EPs were only available on cassette and twelve-inch vinyl. In late 1989 they released them together on one CD under the indicatory title *13 Songs*.

"With Fugazi, how you heard them and experienced their music was an integral part of who they were," writes Jace Clayton in his book *Uproot: Travels in Twenty-First-Century Music and Digital Culture*. "Fugazi made a conscious effort to participate in the networks they wanted to strengthen."[52] They turned down features in *Rolling Stone*, they continue to turn down big money to reunite and headline big music festivals, and they went on an indefinite hiatus in 2003, the same year that MySpace launched. As they sing on "Blueprint," from *Repeater*: "Never mind what's been selling. It's what you're buying."

No set lists meant the band had to be prepared to play any of their songs at any show, a feat that would flummox most other bands, the loosest of which still arrange an impromptu list of songs to play prior to going on stage. Leveraging not only the spontaneity and variety of their live sets but also the low overhead of the downloadable MP3, Fugazi archived and later offered individual recordings of all of them for download on the Dischord site for five dollars apiece. The site states, "Between 1987 and 2003, Fugazi played over 1000 concerts in all 50 states and all over the world. Over 800 of these shows were recorded by the band's sound engineers. The goal of this project is to make each of these recordings available to download for a small fee. A searchable archive lists each show along with available photos, flyers and general show info."[53] MacKaye's recent explanation of his own archival impulse is worth quoting at length:

> I build archives precisely because I have a sense of custodial responsibility. A lot of my work has been focused on the idea that not only can you build your own road, but you can drive on it too. But the problem with these small roads is that they get built off to the side of super highways. They tend to be less used, and they atrophy; the vines come over, then people think they're not passable. That they're not possible. But they are passable—they're just not permanent. Super highways are permanent because the people who own the super highways, who erected the tolls, keep them that way. It's just a different track. And it's important that people know there are other possibilities.[54]

Punk archivist Ian MacKaye at the Dischord House in Washington, D.C.
Photo by Pat Graham. © Pat Graham Photography. Courtesy of Pat Graham.

MacKaye's metaphor is similar to a concept in hip-hop regarding your place in the culture that better suits the current state of media at large. To avoid conflict in hip-hop it is said, one creates and stays in one's own lane. No merging, no switching, everyone runs parallel. Our larger media culture is much more like a superhighway now than it is a flowing stream of any sort. Everyone—creator and consumer—finds or makes their lane and stays in it, ever enabled by personal media, mobile devices with networked screens.

"You have to remember that this is why Dischord Records was started," MacKaye says seriously when I mention the MySpace server migration. "Not because we wanted to have a record label, but because we wanted to document something that was important to us. We didn't think the world needed to have a Teen Idles record. *We* wanted the Teen Idles record. It was important *to us*. So, things that were important to us, we hung onto, and we continue to hang onto."[55] MacKaye's mother was an archivist in the most personal sense. She kept journals for most of her seventy years, tape recordings of the family, and filing cabinets of correspondence and genealogical research. So his archival impulse is also a historical act. Outside of large institutions like the Rock and Roll Hall of Fame or the Grammy Awards and the corporate major labels—melted master tapes notwithstanding—no one is going to keep a record of the things we think are important if we don't. "They own the history," MacKaye says. "Knowing that, I feel like it's important to hang on to evidence of prior civilization, the pottery shards that let people know that they weren't the first."[56] His words sting, since we're now living in what might be the most recorded yet least remembered era of humanity.

"I feel really clear that the work that I've done—that we've done—was about kids doing something they wanted to do and showing that it's possible despite what the corporations say," MacKaye explains elsewhere. "Leaving markers, or breadcrumbs, so that people know this is a possibility—I hope that inspires other people who inevitably will come along to do the same."[57] The punk practices of DIY culture may be a cul-de-sac, but they're always available—and in ever-evolving forms.

BITTERSWEET DISTRACTORS

If Radiohead has been plagued with anything throughout their career, it is expectations. Once the single from their first record, "Creep," broke in 1993 and became an MTV staple, the band was tainted for a time with the "one-hit wonder" stigma. Their more-focused, more-consistent, and just plain more-rocking 1995 follow-up, *The Bends*, certainly saved them from that, but could they top it? They could and did: the 1997 *OK Computer* is one of the most lauded and loved albums of the 1990s. Without a hint of hyperbole, many said it "saved rock and roll." Not bad for a band named after an old Talking Heads song.

In lieu of traditional promotion, their next album, 2000's *Kid A*, streamed online for weeks before its release. Radiohead had insisted on no lead singles, no music videos, and no interviews. Robin Sloan Bechtel, the head of new media for Capitol Records, had been exploring the internet's promotional powers throughout the 1990s. She saw *Kid A*'s release as a way to legitimize the internet to her bosses. "Everything in the industry at that point was like, 'The Internet isn't important. It's not selling records'—everything for them had to translate to a sale," Bechtel said in 2015. "I knew the Internet was [generating sales], but I couldn't prove it because every record had MTV and radio with it."[58] Bechtel had done a screensaver for the Beastie Boys, created a website for Megadeth, and launched the first online single debut for Duran Duran. Now she wanted to stream Radiohead's next record in its entirety weeks before it came out.

Outside of an all-encompassing indifference to the online world, Capitol's main worry was piracy. "We put up the whole album three weeks before the street date—this was unheard of," Bechtel says. "I don't even think the label knew half the stuff we were doing."[59] The album was streamed more than 400,000 times, but on official release *Kid A* was Radiohead's first record to debut at number one on the US charts. It wouldn't be the last time Radiohead used technology ahead of its time.[60]

Overall, the band's *In Rainbows* is somewhere in between the dead-on orchestration of *OK Computer* and the stripped-down eeriness of 2003's *Hail to the Thief*, with a touch of the icy distance of *Kid A*. The songs sound like they were originally composed of much more than is there, then chipped away until all that remains is just what is necessary to stand. While the sound on their first few records swung from traditional rock to experimental electronica, *In Rainbows* seemed to have established a more stable, though still quite eclectic sound. It teeters somewhere between more mellow versions of proggy post-rock and crunchy, dehumanized electronica, but there is always a gooey pop center just waiting to ooze out. That is not to say that Radiohead's signature sprawl is missing—far from it. Most of *In Rainbows* is epic if it is anything, but it's neither the lush instrumentation of *OK Computer* nor the rousing guitars of *The Bends*.

Radiohead released *In Rainbows* as a pay-what-thou-wilt MP3 download. The link stayed up for two months in late 2007, after which they released it on traditional formats. The record went on to platinum sales.[61] When Radiohead decided to release the record on their own, without the help of a record label, others followed. Records by Nine Inch Nails, Oasis, Jamiroquai, and others were offered online shortly after, marking the further shifting of formats widely accepted by the established record industry, but not every band was yet that comfortable trading tradition for technology.

EVERY FORCE EVOLVES A FORM

To create a spike of novelty high enough to be seen by history depends on a lot of things aligning: an open-armed zeitgeist, an open-minded public, a little bit of chaos, and a lot of charisma.[62] Sometimes they become folklore, affecting only those who were there. Points on the map of music history like Woodstock, Altamont, or the June 4, 1976, Sex Pistols show in Manchester, where supposedly everyone in attendance left that show dead set on starting a band.[63] Other times these events are recorded, as great performances, works of art, books, or records.

"It has become increasingly clear that 1979–80 [. . .] was a threshold moment," Mark Fisher wrote, "the time when a whole world (social democratic, Fordist, industrial) became obsolete, and the contours of a new world (neoliberal, consumerist, informatic) began to show themselves."[64] It was the beginning of post-punk. Art school kids were shaping the shards of punk rock into ever new sounds. As one of the more significant of these, Gang of Four combined the lean muscle of punk with the bare bones of funk. As cofounder Andy Gill once said, "You could tell by listening to Gang of Four music that punk had happened."[65] Lyrically social and political, their lanky limbs swung hard and wide against the "middle-class malaise" of the 1970s.[66]

The first time I heard Gang of Four's *Entertainment!*, suddenly much of what I was already listening to made much more sense. Fugazi had a lineage. Naked Raygun had context. Wire had contemporaries. During the post-Lollapalooza package tour phase of the 1990s, I finally saw them live in 1991. It was a woefully crippled lineup that only included Gill from the original Four, sharing the stage at Atlanta's Fox Theatre with a motley mess of bands: Young Black Teenagers, Warrior Soul, Public Enemy, and the Sisters of Mercy. The fact that Gang of Four was considered viable in that lineup ten years past their prime is significant, though.

Woven as an influence or wielded as an instrument, *Entertainment!* remains a relevant aspect of modern music's raw materials. Frank Ocean sampled "An-

Dave Allen performing with the reunited Gang of Four in 2005. Photo by Heidi Kirkpatrick. © Heidi Kirkpatrick. Courtesy of Heidi Kirkpatrick.

thrax" for the song "Futura Free" on his 2016 record *Blond*, and El-P sampled "Ether" for "The Ground Below" from the 2020 Run the Jewels' record, *RTJ4*. It was number eighty-one on *Rolling Stone* magazine's 2013 "100 Best Debut Albums of All Time" list, and in 2012, when they updated their 2003 list of the "500 Greatest Albums of All Time," *Entertainment!* moved up from 490 to 483, a seven-spot jump in a decade, over forty years after the record was released. It stands at number eight on *Pitchfork*'s "Top 100 Albums of the 1970s" list for 2004. Even 2021's retrospective box set represents their earliest era: *Gang of Four 77–81.*

So, when the original four reformed in 2004, as if to prove how strident those early records were, they rerecorded those classic songs. The result was *Return the Gift*, which predominantly features tracks from *Entertainment!* and its follow-up, *Solid Gold*, performed live on a soundstage. By the time they released *Return the Gift* in 2005, there were bands that had drawn direct influences from the original Gang of Four, and there was a whole new generation of writers and fans interested in the post-punk they had helped launch, and they were comparing a new crop of bands to them.[67]

"Those bands helped us get back into the limelight with a whole new generation of music fans," said original bass player Dave Allen, "who came along thinking they were going hear Bloc Party or Franz Ferdinand and then got their minds shattered." Though they are often considered overtly political, Allen bristles at the connotation. "People would say, 'Rage Against the Machine is just like Gang of Four.' As much as I respect those guys and what they do, our aims were very different. We weren't revolutionaries. We were dissecting everyday life."[68] They were political but only in the most personal sense.

After touring with the original lineup, Jon King and Andy Gill were ready to record a new album, but Hugo Burnham and Allen didn't think the world needed a new Gang of Four release. Allen, having spent many intervening years consulting bands on negotiating the music industry's new digital landscape, wanted to do something new, something different. He told me at the time, "If we don't own the idea, there's no point in doing it."[69] He told *Willamette Week* in 2014,

> What I'd wanted to do instead was set up cameras in our rehearsal room in London and do what Radiohead did. This would have been a perfect Gang of Four moment: You can check in on our working methods, you can check in on the arguments that take place. You'd get the chemistry of the band, and then I just felt like, let the crowd decide: What do you think is worth following up on? We'd still never make an album, just complete these songs, and leave them up on YouTube so millions of people could stream them forever, and you don't have to pay a thing. Meanwhile, our cachet goes up in the world for touring, and we can go out again. That's what the Web's for. In music, I think the Web gives you this massive distribution system out of the hands of radio, out of the hands of distributors, out of the hands of record labels. What could be better for rock 'n' roll than that?[70]

This sense of independence, the lingering influence of punk inherent in postpunk, translates well on the Web at a certain scale. It becomes both participatory and tactical, privileging performance over product.[71]

TORN TOGETHER

If you were familiar with the work of Jaime Meline and Mike Render before they got together, they seem like a most unlikely team. Once you hear them together, their pairing makes so much sense.

Meline is a rapper and producer better known as El-P, who emerged from Brooklyn in the mid-1990s as a part of the New York underground scene. El-P was the name, and Company Flow was the group. If you were heavily into hip-hop in 1997, you were familiar with *Funcrusher Plus*. As Company Flow, El-P, Big Juss, and Mr. Len blazed onto the scene in sharp contrast to the shiny suits and "Big Willie" style of what was popular at the time. Noisy and stuttering, theirs was a gritty groove that sounded as much like something recently unearthed as it did a future yet to come. It was decidedly more dusty than digital.

After Company Flow split, El-P went on to start his own label, Def Jux, releasing influential records by Cannibal Ox, Aesop Rock, RJD2, Hangar 18, and his own solo records. "I was born and raised in New York City and grew up on some ill, B-boy shit," he told me when his second solo record, *I'll Sleep When You're Dead*, was released in 2007. "So, this is me. Everything that emanates from me is an extension of that—it's built in."[72] As deeply rooted as it is in hip-hop orthodoxy, that essence has evolved with the times. No one would say that El's solo material sounds like Company Flow, but you can hear the history. "I believe in reference, but I don't believe in imitation," El says. "I don't hold on to too much nostalgia because I don't have to."[73]

Mike Render, a.k.a. Killer Mike, is a second-generation member of Atlanta's Dungeon Family and made his debut with a standout guest verse on OutKast's 2000 record, *Stankonia*. Through major labels pushing his release dates back, leading him to release his records himself, to signing with T.I.'s Grand Hustle imprint, Mike has seen the ills of the industry as clearly as he has seen the ills of society. As outspoken as an activist as he is as an emcee, he famously updated Martin Luther King Jr. in 2015 saying, "riots are the language of the unheard."[74] Mike tapped El to produce his fifth record, 2012's *R.A.P. Music*, a project that made promises they would soon deliver on.

Since 2013 they have been performing together as Run the Jewels.

El-P and Killer Mike are Run the Jewels. Photo by Timothy Saccenti. ©
Timothy Saccenti. Courtesy of Timothy Saccenti.

Through a string of posts online, the duo announced their arrival and that
their debut would be released as a free download.[75] As discussed throughout
this book, the technologies of recording and playback have evolved irrevocably over the decades these two artists have been making music. From hustling
vinyl and mixtapes to watching the internet tank CD sales, they've experienced
the changes as artists as well as entrepreneurs. So, when they got together and
formed Run the Jewels, they were determined to give away the music online.
They weren't the first to do such a thing, but it was a bold move regardless.
After what they had witnessed with labels and releases over two decades, they
decided giving the music away was the only way.

By the time Run the Jewels released their first record, streaming had become
the primary distribution channel for music, which was fine by them. "We don't
have anything to offer except this music," El says. "We don't have a big budget,
don't have videos or all that shit. And I couldn't think of a bigger gesture to the
world and people who'd be interested in this music, than to say—the only thing
I have, I'm giving to you. As a down payment on what I hope is a relationship.
On what I hope will be you falling in love with this, and then somehow that

working out so that I get to do what I love to do."[76] After his experience with piracy as a label owner, El defended the fans as well. "The reason people were downloading music, is because they love music, remember? It's not because they're criminals. I don't think it was radical. [. . .] It just felt right."[77]

Music has been used as mere marketing for a long time now. From the idea of using one hot single to sell an eighteen-dollar CD to using music videos to launch streetwear and drink brands, music has been a doorway into people's lives. It shows up in movie soundtracks, as ringtones, and licensed for commercials and TV shows, but the richest recording artists aren't rich off their music. Their music lent them access to other money-making ventures. If you look at the nouveau riche of music, artists like Jay-Z, Beyoncé, Kanye West, 50 Cent, or Rihanna, you'll find that music is only a fraction of their wealth, a pretty face on their massive fortunes.

None of that is to say that Run the Jewels music is just marketing, but giving it away was a risk the duo was able to capitalize on later with tours, merchandise, and physical formats. "Maybe the industry won't be able to support a lower tier, the way it used to," El continues. "But when that happens, the lower tier finds its own route."[78]

As we thin out and streamline every form of media interaction through likes and shares and comments, the design of the media, the bases of the algorithms, and the infrastructure of the devices and networks play a role that we cannot ignore. Now that everything has gone not only digital but also mobile, whatever it is you're pumping into your head via that black box in your pocket, whatever is occupying your attention, its roots run deep in separate devices, in the cultural components of eras long passed.

PART II
THRESHOLD

Walls, Windows, Doors

What is my point of threshold? What does it mean to dance on a public stage? Everything is getting vague. But just because I am constantly taking into account that I am overcoming a threshold, is this a cartographic initiative, an attempt at localizing myself within an environment? Or am I realizing that I am an environment as such?
—Min Tanaka, "Body-Assemblage"

The transitional periods have all the properties of the threshold, the boundary between two spaces, where the antagonistic principles confront one another, and the world is reversed. . . . The rites associated with these moments also obey the principle of maximization of magical profit.
—Pierre Bourdieu, *The Logic of Practice*

A threshold is a sacred thing.
—Porphyrus

PART II

THRESHOLD

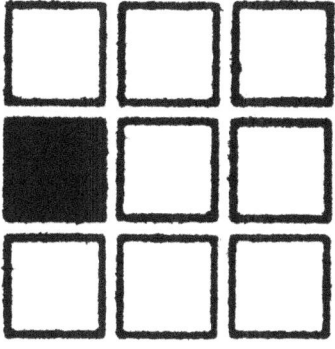

Time of the Signs

We started picking them before they could ripen. A certain crucial
growing period was lost, as marketing evolved and the mechanisms
of recommodification became quicker, more rapacious. Authentic
subcultures required backwaters, and time, and there are no more
backwaters. They went the way of geography in general.
—Cody Harwood, in William Gibson's *All Tomorrow's Parties*

It's funny how the colors of the real world only seem really real
when you viddy them on the screen.
—Alex, in Anthony Burgess's *A Clockwork Orange*

Why Am I always looking at life through a window?
—Charlie Gordon, in Daniel Keyes's *Flowers for Algernon*

GENERATION X WAS A BAND

An era can be said to end when its basic illusions have been exhausted.
—Arthur Miller

Before it was the designation of people born between 1965 and 1980, *Genera-tion X* was a book. Published in 1964, Jane Deverson and Charles Hamblett's volume is full of interviews that highlight the disparities between the dead-end attitude of British youth of the era with the "the cozy images projected by the advertisers and moralizers, packages of false glamour and journalistic optimism."[1] The existence of this artifact suggests that the generation marked with an X started earlier than is generally accepted. Bass player Tony James found a copy of the book under soon-to-be pop icon Billy Idol's bed in 1976 and decided that it was to be the name of their band. The band Generation X was appropriately short-lived and had minimal impact outside of the London punk scene. When Idol went solo in the early 1980s, James went on to form Sigue Sigue Sputnik.

If you were looking to get CliffsNotes for the 1980s in musical form, you'd be hard-pressed to do better than Sigue Sigue Sputnik's 1986 debut, *Flaunt It!* Tony James described them as "hi-tech sex, designer violence and the fifth generation of rock and roll."[2] A product of punk in the same way that Big Audio Dynamite and Devo were, their techno-pop sound, driven by double electronic drummers, was laced with samples from movies and media. Even after all of the work Trevor Horn had done defining a new sound for the decade, Sigue Sigue Sputnik was still exciting. With skyscraping spiked hair and ripped-up glam gear, they had a space-trash look, like leftovers from the ends of an alternate timeline. For better or worse, they looked and sounded like the future.

In a move of unfortunate prescience, the band sold brief advertisements that played between the songs on the record. Real ones for *i-D* magazine and Studio Line from L'Oréal shared space with fake ones for the Sputnik Corporation and a Sputnik video game that never materialized. James touted the spots as commercial honesty, adding, "our records sounded like adverts anyway."[3] Like the punk that preceded them, Sputnik were destined to implode.[4] Yet whereas punk railed against the dominant culture, Sputnik was out to mirror it, to con-sume it, to corrode it from the inside. The former was trying to kill it; the latter was trying to eat it alive.

Chris Kavanagh, Tony James, and Martin Degville (*left to right*) from Sigue Sigue Sputnik live at Abbey Road Studios, London, November 29, 1985. Licensed from dpa picture alliance / Alamy Stock Photo.

It would take Billy Idol nearly a decade to embrace technology in the cyberpunk fashion that Sigue Sigue Sputnik had done in the early 1980s. The reaction to Idol's 1993 concept record, simply titled *Cyberpunk*, is perhaps the best example of two of the attitudes competing inside Gen X. The Information Superhighway, as it was often called at the time, was just making inroads into homes around the world, and its technology-based, cyberpunk, DIY influence was also creeping into the culture at large. In his 1996 book *Escape Velocity: Cyberculture at the End of the Century*, Mark Dery describes Idol's *Cyberpunk* as "a bald-faced appropriation of every cyberpunk cliché that wasn't nailed down."[5] In contrast, our friends and O.G. cyberpunks Gareth Branwyn and Mark Frauenfelder advised Idol and were involved in various aspects of the record's release.

The equally ridiculed and celebrated cynicism of Generation X was born of love. It sprang from our underdog status, and it persists from watching so many tightly held joys disappear. It's not a product of nihilistic disdain, uncontrollable ennui, or generational animosity. It's a product of watching what you love die, disappear, or be commodified beyond recognition. The loss of joys happens to every generation, but our underdog status is unique. We are the first generation to earn less money than our parents *and* our children.[6]

One of the unspoken yet central tensions among Generation X entails the idea of cultural ownership. It is the old battle of the underground versus the mainstream but also the desire to introduce one to the other, to be the one who shepherds something from subcultural obscurity to mainstream success. It is the need to subvert the clichés of the previous generation but also to be successful in spite of them, as embodied by an act like Sigue Sigue Sputnik. It is a realization that you can't change the world, so you might as well try to succeed in spite of it.[7] Coming of age in the 1990s, Gen X might be the last generation to feel these contradictions of capitalism.[8] We might be the last generation for whom the concepts of *underground* and *mainstream* have any real meaning. The terms are still in use, but, as we will see, they don't denote the divisions of market share they once did.[9]

Also straddling this cultural shift, LL Cool J came to fame in the mid-1980s with the Rick Rubin–produced records *Radio* in 1985 and *Bigger and Deffer* in 1987. The scrappy rawness of the young Cool J's raps over Rubin's reductive production proved irresistible to both the streets and the charts. By the time he released *Walking with a Panther* in 1989, Cool J was rich. Though the record sold well, it also suffered a severe backlash. The Gen X–led hip-hop community of the time was nonplussed by the overt materialism. Though not a complete conceptual failure, the record was way too long and unabashedly reveled in its own excesses. At the time, Cool J's posturing came off as posing. Ten years later, during the Big Willie era of the late 1990s, conspicuous consumption was one of the prevailing modes of popular hip-hop. From Nas to Biggie to Jay-Z, the contradictions of capitalism were on display, and only the orthodox of the underground were complaining. The underground was supposed to stay buried. That's what Darby Crash meant when he said, "What we do is secret!"[10] Cool doesn't scale.

Sigue Sigue Sputnik's slogan was "fleece the world," and their punk predecessors the Sex Pistols, while they railed against it all in 1976, reunited twenty years later, citing "your money" as the sole reason. These ideas were still somewhat shocking then, but now they seem downright quaint. The author of 1991's *Generation X: Tales for an Accelerated Culture*, Douglas Coupland recently reassessed the generation, writing, "Today, I wonder how much of what we call a generation is simply a matter of any given temporal cohort's tech exposure during their pre-pubescent neural wiring."[11] As we are each imprinted by the media we have grown up with, our respective generations cohere less around our birthdays and more around collective nostalgia and native technologies.[12]

ADVENT HORIZON

Technology is anything that was invented after you were born.
—Alan Kay

The commonly accepted span of a generation is twenty to thirty years.[13] Right now, there are more generations alive at the same time than ever before.[14] In his book *Human as Media*, Andrey Miroshnichenko describes the phenomenon in terms of eras, writing, "If an era is shorter than a generation, the balance between the speed of technological innovation and the speed of cultural adaptation breaks down."[15] We feel at home in the era we're born into. We are familiar with its media and adept with its technology. We feel a sense of loss when we cross from one era to another.

In the Socratic shift from speaking to writing, and in the transition from writing to typing, humans have been comfortable—differently on individual and collective levels—in one of the phases. As we adopt and assimilate new devices, our horizon of comfort drifts further out while our media vocabulary increases. I have never known a world without television, and my students have never known—or don't remember—a world without computers, the Web, or cell phones.[16] This is an example of what I have called the *advent horizon*. It takes thirty years for a full, generational change and with that a full shift in advent horizons. The ubiquity of personal media devices enables what writer Michael Harris calls "the end of absence." In his book of the same name, he notes, "If you were born before 1985, then you know what life is like both with the internet and without. You are making the pilgrimage from Before to After."[17] The internet is the epochal shift of our time, and few and fewer of us remember how life was before it.

Evidence that we have crossed one of these lines isn't difficult to find.[18] For example, you can see it in the resurgence of vinyl record sales since 2010. The older format started outselling the CD in early 2023 for the first time since the 1980s.[19] Even Christian Marclay has long since stopped using vinyl records in his performances. As music has moved on to digital formats and beyond, records have been recontextualized since he started using them in the 1980s. Once they were no longer the dominant format, DJing, even in the manner that Marclay does it, meant something different. He told David Toop in 2011, "Now it's retro. Culturally it has a different signification."[20] Older fans of vinyl records are clinging to their youth, celebrating the only music format that they feel mattered, while younger ones are longing for a time they never experienced.

Here's another example: I visited a major art and design university several years ago. In their animation and game design curriculum, students take illustration (with pencils and paper), flipbook-style animation (with paper and lightboxes), and 3-D modeling (real-world 3-D, sculpture with clay and other materials) before they ever sit down at a computer. Though these students are natives to the latter, they are forced to learn their skills in the analog world before moving to the digital.

Similarly, there's a fundamental, experiential difference between browsing actual books on physical shelves or records in racks and browsing their titles online or in a database. The two experiences are not only dissimilar, they are also related in only the most tenuous way. A book or a vinyl record is an *analog totem* from a previous era.[21] Teaching illustration and animation on paper before moving to computers is *analog scaffolding* for the digital world. Clinging to a previous era and having to back up to learn something new: these are both evidence of a new era, that an advent horizon has been crossed. Each generation is born during a certain technological era, between these lines we draw.[22] We are imprinted by the media technology with which we grow up.[23]

Perhaps our children will cross a line of comfort when something new becomes the norm for their children, but the world before wireless connectivity means nothing to them. Every generation witnesses vast changes wrought by the technology during their time. Generation X is the last to remember daily life without the internet and its mobile appendage the cell phone. Pop culture extends the second childhood after adolescence further and further into adulthood. From punk and hip-hop to skateboarding and gaming, so many popular interests and causes are tied to youth. The conflicts between growing up, growing old, and staying true to ourselves are more and more evident in twenty-first-century Western culture. Our memories are fallible and ever-more mediated, yet they are important to study. "They tell us about the ways in which people construct the past," writes Mary Fogarty in her study of aging breakdancers, "and within this practice they reveal the value systems highlighted by different generations."[24] We construct and cling to pasts that our present can never live up to.

Part of the problem is cognitive. Our brains' ability to create and store new memories slows down, to a near-stop, therefore making our most cherished memories those of our youth. And when we remember those times, we reify them, making them stronger. So, being stuck in the past is basically a somewhat natural state for our brains, and our technology lets it linger longer than ever. Another part of the problem is cultural. We define ourselves according to our cohorts, tied to both time and tribe. As John Naisbitt pondered in 1994, "The riddle of the 1990s is, what's going to become universal, and what's going to remain tribal?"[25]

NOMADS AND MONADS

Walls are perfect objects to play things on, scenes which join together
the sense of "beyond" and the sense of confinement.
—Reza Negarestani, *Cyclonopedia*

Your screen my screen we all screen.
—The Kid, in Cormac McCarthy's *The Passenger*

You've been in this room: You're home for the holidays. Looking around the living room at parents and siblings, you notice that most of them are pecking away at their smartphones, two are also wearing headphones, and one is clicking on her laptop. The television is on, but no one's watching it. Each person is engrossed in their own solipsistic experience, be it a game, a TV show, or some social medium. The habits we inherit from these devices might be unintended, but it is up to us to resist them. Such thresholds are the beginnings of new levels of being.[26] Our new state seems to be experiencing the world as individuals.

I started teaching college as a graduate student in 2002. In the time since, I have seen various devices infiltrate the classroom: from no computers or phones in the room to special rooms equipped with computers and projectors (called "smart rooms"), and finally to every classroom fully equipped with computers and screens and every student with their own laptop, tablet, phone, and other various gadgets.[27] Mark Fisher described this screen-enabled environment as "the space in which humans now primarily live, and one that is both shaped by, and manufactured from, their desires and drives."[28] As we get lost in these tiny, liminal worlds, we risk losing ourselves in the larger one.

We gather around some screens, the likes of large flat-screen displays connected to projectors, and use others—laptops, tablets, and phones—individually. But affordances are not merely infrastructural. They are also behavioral. That is, just because technologies are capable of many things doesn't mean we use them for all of those things. Moreover, we often use them for purposes for which they were never intended. Many affordances emerge from use.[29]

In our evolution from television screens to computer screens and to mobile screens, we've fundamentally changed the infrastructure by which ideas spread. We gather together around big screens to watch passively while, paradoxically, we engage as individuals with smaller screens to connect with each other.[30] Screens at the large scale (e.g., theater and television) turn us into *nomads*, hunting and gathering in groups for news, information, and entertainment.[31] Smaller screens

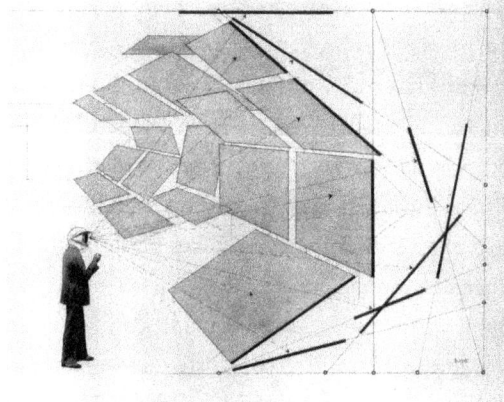

Herbert Bayer, *Diagram of the Field of Vision*, 1930. © 2024 Artists Rights Society (ARS), New York / VG Bild-Kunst, Bonn.

turn us into *monads*, experiencing media individually or together in what Michael Bull calls "accompanied solitude."[32] Feeling disconnected from the world we have created is a natural response to the fracturing of mediated experiences. Music, for example, once listened to with others, is now stowed away in portable black boxes and usually experienced alone.[33] Other forms of media are delivered person-to-person via invisibly connected screens.

The disconnections inherent in our slicing up the world with media operate at different scales and via different channels. Rosi Braidotti writes, "The nomad and the cartographer proceed hand in hand because they share a situational need—except the nomad knows how to read invisible maps, or maps written on the wind, on the sand and stones, in the flora."[34] Writing before the spread of personal media, before these technologies found their way into our pockets, Marshall McLuhan postulated "electronic-age" nomads. "[People] are suddenly nomadic gatherers of knowledge, nomadic as never before, informed as never before, free from fragmentary specialism as never before—but also involved in the total social process as never before; since with electricity we extend our central nervous system globally, instantly interrelating every human experience."[35] Our media networks have since gone global while their attendant physical devices have gone mobile. Screens of different sizes and mobility shape our world in distinct ways. That is not to say that one can't go to the theater or watch television alone, but the design of these media lends them to sharing, whereas the mobile screen is intended for the individual. That is the price of being a monad.

With screens that brought the world inside the walls of the home, the barrier between the public and the private was broken. Making sense of the distinction when it comes to the mobile screen is a problem. The telephone, once considered a private matter between two callers, is now often shared in public spaces, or tapped, recorded, and stored for later scrutiny. Like the television or theater, it can be used in groups, as the boardroom conference call illustrates, but it tends toward solitary use, even if that use often takes place in public places. The telephone, as it has become mobile, has succumbed to the screen, so much so that a cell phone without a full-color touchscreen quickly became an antiquated concept. The window to the Web and to the world that was once in the domain of the office or living room converged with every other media form on the computer desktop and is now in pockets and bags dispersed throughout the human world. Given their ubiquity, it isn't difficult to see how seductive the screen can be, but the computer did not usher in the information age.[36]

The transition from page to screen has been privileged by those who study these changes, but the shift to broadcast media was also prefigured by the lecture circuit and stage plays. Predating McLuhan, the Russian filmmaker and theorist Sergei Eisenstein wrote in 1923, "The spectator himself constitutes the basic material of the theatre."[37] We are easily lost in both words on the page and drama on the stage. In his *Confessions*, Saint Augustine lamented straying into fantasy, writing, "Stage-plays . . . carried me away, full of images of my miseries, and of fuel to my fire. Why is it, that man desires to be made sad, beholding doleful and tragical things, which yet himself would by no means suffer? yet he desires as a spectator to feel sorrow at them, and this very sorrow is his pleasure. What is this but a miserable madness? for a man is the more affected with these actions, the less free he is from such affections."[38] Augustine's confessions of losing himself in plays are all too familiar to us. Augustine continues on the same topic: "What marvel that an unhappy sheep, straying from Thy Flock, and impatient of Thy keeping, I became infected with a foul disease? And hence the love of griefs; not such as should sink deep into me; for I loved not to suffer, what I loved to look on; but such as upon hearing their fictions should lightly scratch the surface; upon which, as on envenomed nails, followed inflamed swelling, impostumes, and a putrefied sore. My life being such, was it life, O my God?"[39]

The connection Augustine felt with the histrionics of the stage is also a separation from his everyday life, and more importantly for him, from his God. His connection is the same one we feel with movies, television shows, and video games, and the personalities we connect with through them.[40] His worries are the same ones we may feel when we sense ourselves drowning in these media.

FIFTEEN MINUTES IN THE FUTURE

Fame, fortune, and influence are inherently asymmetrical. They all require a one-to-many style of distribution akin to the wide-range broadcasting model of legacy media. As that media infrastructure has given way to smaller and smaller platforms serving smaller and smaller audiences, the ideas of fame, fortune, and influence have been reconfigured. The dictum, "In the future, everyone will be famous for fifteen minutes," for which several sources have claimed credit, is widely attributed to Andy Warhol.[41] Regardless of who first said it, those fifteen minutes of the future are the popular origins of "the long tail of fame." Though the phrase has been around since the late 1960s, its proposed future is here. In the first issue of cyberpunk magazine *Mondo 2000* from late 1989, editor-in-chief R. U. Sirius declared, "In the evolving subculture of graffiti art, interactive video, garage bands, cassette culture, cyberpunks, and fanzines, we find a future where everyone gets to be famous to 15 people. As the technology gets cheaper, smaller, and smarter, pop culture becomes more fleeting and ex-foliative. There's room for everyone to strut their stuff. Everybody gets to play at art and image, text and subtext. Everybody gets to be a little bit Warhol."[42]

In his 1991 essay "Pop Stars? Nein Danke!" Scottish recording artist Momus updated Warhol's supposed phrase and added, "We now have a democratic technology, a technology which can help us all to produce and consume the new, 'unpopular' pop musics, each perfectly customised to our elective cults."[43] As consumers we have found entertainment that feels more and more like it's just for us. As creators we have found audiences that appreciate our time and talents. In *Small Pieces Loosely Joined*, David Weinberger's 2002 book about the Web, he notes about bloggers, content creators, comment posters, and pod-casters: "They are famous. They are celebrities. But only within a circle of a few hundred people." He goes on to say that in the ever-splintering future, they will be famous to ever-fewer people, and—echoing Sirius and Momus above—that in the future provided by the internet, everyone will be famous to fifteen peo-

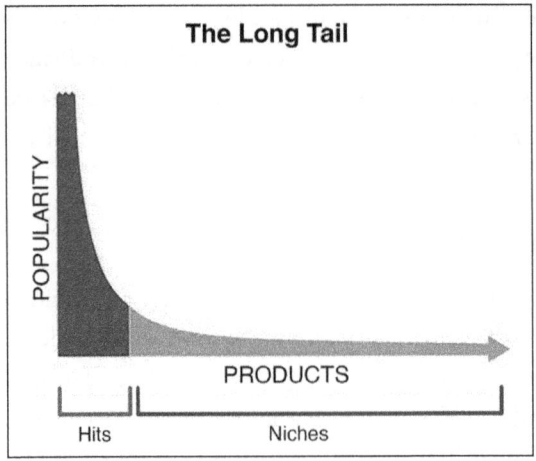

The Long Tail chart.

ple.[44] Democratizing the medium means a dwindling of the fame and notoriety that the medium can support.

Around the turn of the millennium, the long tail, the internet-enabled power law that allows for millions of products to be sold regardless of shelf space, reconfigured not only how culture is consumed but also how it is created.[45] It has since gotten so long and so thick that there is not much left in the big head, and the rare hits are fewer and fewer and further in between. As the online market supports a wider and wider variety of cultural artifacts with increasingly less depth of interest, they each serve ever-smaller audiences. Even when a hit garners widespread attention, there are still ever more of us farther down the tail, each in our own little world. The lengthening and thickening of the long tail plot our media culture as it moves from the shared center to the individuals on the edges, from one big story to infinite smaller ones. Now what does such splintering do to the economics of creating culture?

Bruce Nolan, played by Jim Carrey in the 2003 movie *Bruce Almighty*, is a man unimpressed with the way God is handling human affairs. In response, God lets him have a shot at it. One of the many aspects of the job that Bruce quickly mishandles is answering prayers. His head is flooded with so many, he can't even think. He sets up an email system to handle the flow, but the influx overwhelms his inbox. As a solution to that, he implements an autoresponder to send back a message that simply reads "Yes" to every request.

Many of the incoming prayers are pleas to win the lottery. His wife's sister Debbie hits it. "There were like 433,000 other winners," his wife Grace explains, "so it only paid out seventeen dollars. Can you believe the odds of that?"[46] Subject to Bruce's automated email system, everyone who asked for a winning ticket got one. That's what you get when you're famous for fifteen people for fifteen minutes. That's what you get when everyone's a winner: about fifteen dollars.

The mainstream isn't the monolith it once was. It's a relatively small slice of the total culture now, markedly smaller than it was at the end of last century. For better or worse, the internet has democratized the culture-creating and distributing processes we used to privilege (writing, music, comedy, filmmaking, etc.), and it's brought along new forms in its image. The culture of the internet has splintered ever further via social media and mobile devices. Anyone can now create content and be famous for fifteen people for fifteen minutes—and earn fifteen dollars for their efforts.

THE CLUTTER OF POP

Here there is a tabula rasa of indifference. It is like attending a
family gathering and realizing you are old, and the new generation does
not give a fuck about you or your experience.
—Tade Thompson, *Rosewater*

I've encountered this demographic splintering more and more in the classroom as I try to pick examples from pop culture that everyone will recognize. Increasingly, even the biggest shows and movies I bring up leave most of my students out, and whenever I get into the stuff I actually like, I'm greeted with more blank stares than usual. This wasn't the case when I first started teaching at the turn of the millennium, and it's not just the increasing gap between my age and theirs that's making the difference. The big hits from the end of last century have dispersed into a million niches. As the screens get smaller, so does the number of media experiences we share.

"The media," which once was "the mass media," has trickled down from a one-to-many broadcast model to more of a one-to-one, individualized state. If we're all watching a channel on broadcast television, we're all seeing the same shows.[47] If we're all on the same social network, no two of us are seeing the same thing. The limited access to content via broadcast media used to unite us. Now we're only loosely united via the platform, and the platform itself doesn't matter. The long tail used to describe the splintering of artifacts and attention, from blockbuster hits for all to niche interests for each. Now it also describes the apparatus via which we consume those artifacts, the devices that hold our attention. With the further individualization of our media and the adoption of personal media devices, the postmodern promise of individual viewpoints and infinite fragmentation is upon us.

McLuhan framed these media in the Gestalt psychology terms of *figure* and *ground*, the figure being the overt reference and the ground being the invisible referent. "In any gestalt the *ground* is taken for granted and the *figure* receives all the attention," he wrote. "The *ground* is subliminal, and area of effects rather than of causes."[48] The *figure* is the artifact at hand, and the *ground* is its content, the historical context it's indexing.[49] As the figure fragments, the ground we share shrinks.

The medium is only the message at a certain scale, and that scale is diminished.

This fragmentation in the United States has never been more evident than during the last few presidential elections. None of the old tools could provide a picture of what was going on. "It was the last time we could trust the mass media as a reliable shared narrative," declares Chris Riley in his book *After the Mass-Age*.[50] Think about news media. As the nightly network news spread out into twenty-four-hour cable coverage, so did its audience and its intentions. Riley writes that instead of trying to get the majority to watch, each network preferred a dedicated minority: "Now you didn't win the ratings war by being objective; you won by being subjective, by segmenting the audience, not uniting them."[51] And we met them in the middle, seeking out the news that presented the world more the way we wanted to see it than the way it really was. With the further splintering of social media, we choose the news that fits us best. The messages we post, read, and share are just as likely to come from our friends and coworkers as they are journalists at an actual media company.[52] Moreover, through the social urging of these platforms, we all participate in communication without contemplation. We've gone from *having something to say* to *having to say something*.

Not only does the flood of information and the splintering of media diminish our attentions and erode our interpersonal relationships, it also fragments our culture. In his 1996 memoir *A Year with Swollen Appendices*, Brian Eno

proposes the idea of *edge culture*, which is based on this premise: "If you abandon the idea that culture has a single center, and imagine that there is instead
a network of active nodes, which may or may not be included in a particular
journey across the field, you also abandon the idea that those nodes have absolute value. Their value changes according to which story they're included in,
and how prominently."[53]

Each of us tell our own stories, including the cultural artifacts relevant to
the narrative we have chosen. The long tail is an ironic attempt to depict a big
picture that no longer exists. With its emphasis on the individual narrative,
edge culture more accurately illustrates the current, fragmented state of mediated culture. "What I see instead of there being one line," Eno explains in
a lecture from 1992, "[are] many lines, lots of ways of looking at this field of
objects that we call culture. Lines that we may individually choose to change
every day."[54]

In the late 1960s Mark Granovetter was studying how people found jobs. His
1973 article in the *American Journal of Sociology*, "The Strength of Weak Ties,"
states that each person in a close social network is likely to have the same information as everyone else in that network.[55] It's the weak ties to other networks
that lead to the new stuff. That is, weak ties are a more likely source of novel
ideas and information—regarding jobs, mates, and other opportunities—than
strong ones. Granovetter says, "I put the theory of weak ties together from a
number of things. I learned about hydrogen bonding in AP Chemistry in high
school and that image always stuck with me—these weak hydrogen bonds were
holding together huge molecules precisely because they were so weak. That was
still in my head when I started thinking about networks."[56]

Like most of my research interests, I first noticed these thresholds in music. I
was looking at the tapes and CDs I had on hand one day, and I noticed that most
of my favorite bands didn't fit into established genres. They tended to strad-
dle the lines between genres. In nature, these interstitial spaces are called *edge
realms*. In her book *When Plants Dream*, Sophia Rokhlin describes them: "The
edge describes the place where two distinct ecosystems meet. These are places
of tension and unfamiliarity, territories of confrontation, where different ecosystems overlap and merge. The edge is found where a grassland meets a forest,
where oceans reach the shore, where wetlands mediate between river basins and
fields. Edges are hot spots of biodiversity that invite innovation, intermingling,

and new forms of cooperation from various species. Edge realms are thresholds of potential and fecundity."

An edge realm is a wilderness, a mutant space ripe for new forms. In Jeff VanderMeer's Southern Reach trilogy, the mysterious Area X is just such a space. Its pollinations cross well established boundaries, mixing into ever-new breeds, combinations, doubles, and mutations. In his book about VanderMeer's work, *None of This Is Normal*, Ben Robertson writes, "Area X is something else, what has always already disrupted the processes by which borders are established between that and this, between one space or time and another space or time, between the human and whatever its other happens to be."[57] The fertile ground is in between the established crops of others. The new stuff happens at the edges, between the codified categories. Any old boring story from history can be made more interesting by varying viewpoints.

The members and fans of subcultures—groups united by similar goals, practices, and vocabularies—represent what Etienne Wenger calls *communities of practice*.[58] To translate differences and aid communication between these communities, they use what Susan Leigh Star and James Griesemer called *boundary objects*.[59] A boundary object can be a word, concept, metaphor, allusion, artifact, map, app, or other node around which communities organize their overlaps and interconnections. These connective terms emphasize groups' similarities rather than their differences, allowing them to communicate and reduce confusion between their disparate vocabularies. Boundary objects between different communities of practice open borders once inaccessible, circulating ideas into new territories. Allusions, references, quotations, metaphors, and other figurative expressions provide the points at which multiple texts, genres, and groups connect and collaborate. Hunting and gathering, picking and choosing, we can each make our own edge culture.

Mark Granovetter conceived the edge realms of these cultural networks way before we were all connected online, but his insight is all the more relevant today. "Your weak ties connect you to networks that are outside of your own circle," he explains. "They give you information and ideas that you otherwise would not have gotten."[60] With our personal media, ubiquitous screens, and invisible, wireless networks, we live in a world of weak ties.

"We live in a time I think not of mainstream," John Cage said in 1992, "but of many streams, or even, if you insist, upon a river of time, that we have come

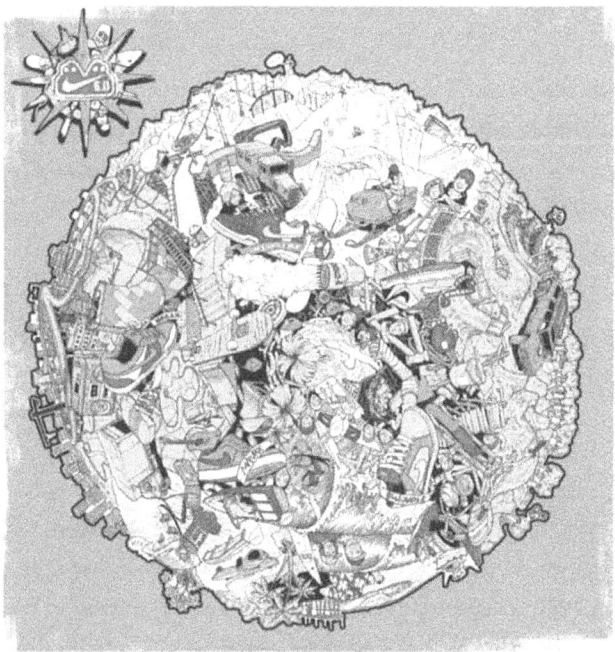

One world, one market. Illustration by Adam Haynes for
Nike 6.0. © Adam Haynes. Courtesy of Adam Haynes.

to delta, maybe even beyond delta to an ocean which is going back to the
skies."[61] None of the top-fifty best-selling records of all time were recorded this
century.[62] Where certain cultural artifacts used to define an era, now there are
none.[63] The infrastructure of the media and our use of them no longer support
a mainstream. The mainstream metaphor isn't even applicable anymore. It implies groundwater, springs, and wells. It implies that there are rivers, rivulets,
tributaries, and flows. It implies the same water, all connected and flowing.

In the 1990s, events like the X Games and the Warped Tour and short-lived
websites like Hardcloud.com and Pie.com tried to gather long-tail markets that
were too small by themselves into viable mass markets, like a sort of cultural
junk bond. It happened with the many musical subgenres of the time, working
the ends against the middle in trying to get the best of both worlds, ending up
instead with the worst of them.[64] The mono-brow mixing of high culture's concerns with low culture's lack thereof only makes sense if there's a market in the
middle.[65] What was the label "alternative" if not a feeble attempt at garnering

enough support for separate markets under one tenuous banner? If you can get the kids *and* their parents invested, you might have a real hit.[66]

In the 2000s Nike tried to sneak into skateboarding again with a quiet purchase of the already established skate-shoe brand Savier, but it wasn't long before they were found out.[67] Then they boldly moved in again with Nike SB and Nike 6.0. By the time of their last attempt, the market for skateboard shoes had given way to a crowd who didn't have the same allegiances to the so-called underground. If their attempt at selling skate shoes in the 1990s was aimed at Generation X, Nike 6.0 was aimed at a younger crowd. They seemed to see that by avoiding or subverting one cliché, you're just creating a new one, so they were unconcerned with the previous generation's ideas about authenticity.[68]

The "6.0" suffix referred to six domains of extreme sports: BMX, skateboarding, snowboarding, wakeboarding, surfing, and motocross. Whatever the practitioners of such sports might share in attitudes or footwear, they don't normally share in an affinity for each other. We ride differently. We dress differently. We think differently. Like the staunch fans of different subgenres of music, we remain in our silos, refusing to cross-pollinate in any way.

GOODS AND SURFACES

If marketing can't bring us together, mass tragedy might. In his 2009 novel *Neuropath*, R. Scott Bakker describes the unifying effect of news of a mass or serial murder, in this case, "The Chiropractor," so named because he removes his victims' spines: "In these days of broadband, it was rare for anything nonpolitical to rise above the disjointed din of millions pursuing millions of different interests. The niche had become all-powerful. The Chiropractor story was a throwback in a sense, a flashback to the day when sitcoms or murders could provide people a common frame of reference, or at least something to talk about when polite questions gave out."[69]

Tragic or heroic stories, news and events that everyone talks about, are called *cultural scripts*.[70] Cultural scripts are the way our fragmented networks coalesce, if even briefly, into unified interests and concerns. Cultural scripts are where the content transcends the platforms, where the layers of networks coalesce into a shared experience.[71]

Photography, that once-stable representation of a slice of time, has given way not only to the motion of film but also to digital representation and manipulation.[72] At their onset, mechanically reproduced images rivaled reality's lived experience and set in motion a questioning of the nature of reality. Once those images began to move, lived experiences took another blow. Iain Chambers writes, "In the uncanny property of the computer to present a 'world picture' we confront the boundary set by the screen, the tinted glass that lies between the apparently concrete world and the simulated one of ethereal lights."[73] The digitization and networked distribution of images has rendered such a concern quaint. The screen is easily the most seductive of technologies.

To that end, we have gone from not only wearing the goggles and gloves of virtual reality systems to using our bodies as input devices via the sensors of videogame interfaces, bringing the machine into the room.[74] Whereas our machines' portability used to be determined by the size of the technology available, the size of our devices is now dictated by the size of our appendages. We can make cell phones and laptops smaller, but then we wouldn't be able to hold them or press their buttons individually. Think about the flat rectangle of your smartphone. Its affordances take little advantage of the dexterity of human hands and fingers. Its interface relies almost solely on our mental abilities. The screen on your smartphone might do anything. Joysticks, keyboards, steering wheels, gearshifts, handlebars, brake levers, even older cellular phones, all take advantage of what hands and fingers can do. They all do fewer things, but they do each of them better. Screens have swallowed up the interactive aspects of almost everything. A lot of the labor of using digital interfaces is cognitive. Tempting the system's seduction through repeated and habitual use recalls Eno's *surrender with risk*.[75] Sure, you have to remember that turning the steering wheel to the left causes the car to veer left, but it also physically shows you. The mapping of these functions requires more and more of our brains than our bodies. Eno adds, "I'm struck by the insidious computer-driven tendency to take things out of the domain of muscular activity and put them into the domain of mental activity."[76] This points out the threshold of our being with our technology.[77]

Whether we view ourselves in nomadic groups or as monadic individuals, our experiences collect on the screen like so much data condensation. The philosopher Michael Heim describes the threshold this way:

Realities are representations continually placed in front of the viewing apparatus of the monad but placed in such a way that the system interprets or represents what is being pictured. The monad sees the pictures of things and knows only what can be pictured. The monad knows through the interface. The interface represents things, simulates them, and preserves them in a format that the monad can manipulate in any number of ways. The monad keeps the presence of things on tap, as it were, making them instantly available and disposable, so that the presence of things is represented or "canned."[78]

The surrogate experiences of monads and nomads make up the diverse disconnections inherent in our use of technologies. Large or small, screens deal in illusions, and, as porous as they seem, their interfaces are still just surfaces.[79]

CHAPTER 5

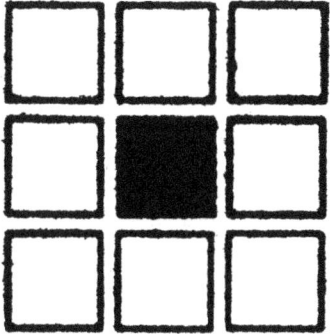

The Surface Industry

Consumption is a system of meaning like a language. . . . Commodities
and objects, like words . . . constitute a global, arbitrary and coherent
system of signs, a cultural system. . . . Marketing, purchasing, sales, the
acquisition of differentiated commodities and objects/signs—all of
these presently constitute our language, a code with which our entire
society communicates and speaks of and to itself.

—Jean Baudrillard, *Simulacra and Simulations*

Media, marketing, and advertising are mechanisms for
abstracting and commercializing codes of human identity
and their molecular components.

—Brian Massumi, *A User's Guide to Capitalism and Schizophrenia*

Do you think a city can control the way people live inside it?"

—The Kid, in Samuel R. Delany's *Dhalgren*

MINING AFFORDANCES

When they engage with this concrete architecture, the
skaters immediately find themselves confronted with the hazards
of solitude. Despite a powerful feeling of belonging, they must
battle first and foremost against themselves to strengthen
their resolve, overcome pain, achieve greater heights and face
the void. Amid the screeching of wheels and the clanging of axles,
each ramp functions like an altar in the celebration of a pagan liturgy.
Again and again . . . It is always after succession of
aborted attempts that the Way becomes clear.

—Joël Vacheron, "Venice Arena"

I don't know any casual skateboarders. Everyone I know who's ever done it has either an era of their lives or their entire essence defined by it—the rebellion, the aggression, the expression inextricably bound up with their being. It's the way you wear your hair and the way you wear your hat. It's the kind of shoes you wear and which foot you put forward. It's the crew you run with and the direction you go. There is something about rolling through the world on a skateboard that changes people forever.

Ever since I first saw Wes Humpston's Dogtown cross on the bottom of a friend's skateboard in the sixth grade, I knew it was going to be a part of my world. I first stepped on a skateboard at the age of eleven. There are scant few physical acts or objects that have had a larger impact on who I am and how I am. Through the wood, the wheels, and the graphics, skateboarding culture introduced me to music, art, and attitude. "Skateboarding is not a hobby," Ian MacKaye says.[1] "We practiced every day. It was a discipline. It was the way that we learned how to navigate our surroundings, and the world was transformed by this navigation because we suddenly saw everything a little bit differently than it had originally appeared."[2] Riding a skateboard fundamentally changes the way you see the world and yourself in it.

Disparaged by pedestrians, police, and business owners, skateboarders reinterpret the urban landscape with style, grace, and aggression. Their instrument of choice, the skateboard, is a humble object of plywood, plastic, and metal. In the 1950s the skateboard, a 2×4 with roller skate frames and wheels nailed to the bottom, was created to emulate surfing during the off season. Carving the curves of empty pools, skateboards and skateboarders eventually moved on to custom ramps and city streets.[3] Riding a skateboard changes how one sees the

The author at age eleven and the beginning
of a very long road.

world generally and the built environment specifically.[4] Where most see streets, sidewalks, curbs, walls, and handrails, skateboarders see a veritable playground of ramps and obstacles to be manipulated and overcome for fun. Hills are for speed. Edges are for grinding or sliding. Anything else is for jumping onto or over. Looking through this lens all one sees are lines to follow and lines to cross. It's a familiar story to skateboarders, the before and after of the world. Skateboarding redefines everyone who does it as well as the way they see the world around them.[5]

The game designer Brian Schrank defines *affordance mining* as a way of determining a technology's "underutilized actionable properties and developing methods of leveraging those properties."[6] An affordance is simply an object's capacity for use, the operations or manipulations it will support.[7] Schrank has mined affordances most famously from computer keyboards, designing games that challenge the use and user of QWERTY keyboards to find new ways of interacting with the common computer. I played a few of his early "mashboard games" when he was still a graduate student at Georgia Tech in 2006. One of them, Hello Zombie Mouth, involved mashing the keys on a keyboard to control a mouth as it yawned and stuttered, trying to say "Hello," urged on by an 8-bit clergyman character. Six regions of the keyboard were each mapped to a phoneme: one to the tip of the tongue, another to the top lip, for example. A few attempts with such a hand-heavy, haptic puzzle quickly demonstrated the untapped potential of such everyday devices for not-so-everyday applications. Such affordance mining can also be found in other underground, DIY cultures like skateboarding and punk.[8]

Like most punk flyers that followed, on October 22, 1938, a young patent office employee named Chester Carlson transferred his handwritten inscription of the date and place ("10.-22.-38 ASTORIA") from one piece of paper to another. Alex Houston describes other shared aspects of the origins of xerography and punk: "from the squalid, nocturnal setting to the noxious odors and subpar gear, the ambience of the Xerox origin story shares an array of signifiers later associated with the grungy clubs and D.I.Y. culture of punk."[9] Grasping its potential for art and appropriation, Andy Warhol had a copy machine installed in the Factory in 1969.[10] With it, Warhol was able to copy and repeat images from pop culture, creating much of the aesthetic he became known for. "By replicating the work many times over," Walter Benjamin wrote at the time in his landmark essay "The Work of Art in the Age of Mechanical Reproduction," "it substitutes a mass existence for a unique existence. And in permitting the reproduction to reach the recipient in his or her own situation, it actualizes that which is reproduced."[11] Benjamin argued that the reproduction of art rid a work of its aura but also democratized its experience.

In her book *Adjusted Margin: Xerography, Art, and Activism in the Late Twentieth Century*, Kate Eichhorn writes, "If copy machines and their gritty output of posters, flyers, and zines helped to define and spread movements intent on bolstering the rights of people on the margins, it was largely against, not with, the grain of the machine's original intentions."[12] One of those movements was punk. Flyers and zines were the primary carriers of information about punk bands and shows. The punk impresario Malcolm McLaren declared that punk flyers "screamed ugliness all across town—designs made to address an army of disaffected youth. These were the rats' ears of the city fighting the consumerist ideology of the mainstream."[13] Lofty ambitions aside, the physicality of these forms is essential to the culture.

"I think it's probably the last youth movement that used paper," Ian MacKaye tells me.[14] Punk, skateboarding, and zines are related pursuits as one documents the other two. "There's a weird relationship between humans and things that are tactile," says the former editor of *Transworld Skateboarding* magazine and *The Skateboard Mag*, Kevin Wilkins. "There's something about the friction of things rubbing up against each other and causing reaction."[15] When zines are mailed out, they're usually accompanied by correspondence of some kind.

"I think paper contains the human gesture in a way that many forms of digital creativity do not," Jenny Toomey tells me. "It almost always involves a greater level of scarcity. You could have the gesture of a human image or a human choice in someone who makes digital art, but it's immediately replicable."[16] Toomey is a towering figure in 1990s punk rock, playing in bands, running a record label, and much more. She and Kristen Thompson, her bandmate in Tsunami and cofounder of Simple Machines Records, put out a zine-style pamphlet called *The Introductory Mechanic's Guide to Putting Out Records, Cassettes, and CDs*, which was repeatedly updated with new and better resources over four editions throughout the 1990s. Toomey continues,

> That's the element of counterculture and punk that I really liked. Not the flashy nihilism of tearing the old down, but rather the joyful enthusiasm of building the new. Asking that increasingly unasked question, "Why are we consenting to so many things we don't agree with? And what would it look like if we tried to build different systems to give us more and better choices?" So, putting out your own records is a piece of it. A nice piece . . . nothing wrong at all with putting out your own records, but it's not going to solve the problem of highly concentrated corporate media systems in late-stage capitalism. It does offer you a way to not completely condone what you abhor.

All of these paper artifacts further foster the hand-to-hand network of indie discourse and the independent sentiment of the subculture, its art, its music, and its rebellion—making its way no matter the obstacles.

SPURIOUS SURFACES

City life is invariably about surfaces, the superficial reading and the
transitory clues involved in our observations of others, hence the overriding
dominance of the visual in urban accounts of experience. The presentation
of self is a largely visual one—the presence of the other is largely a silent
presence in urban culture, even in the urban world of the mobile phone.
—Michael Bull, "The Audio-Visual iPod"

The city, as a form of the body politic, responds to new pressures and
irritations by resourceful new extensions always in the effort to exert staying
power, constancy, equilibrium, and homeostasis.
—Marshall McLuhan, *Understanding Media*

Skateboarders find their own use for everything in the city.[17] First it was surfing the open waves of sidewalks and streets.[18] Then the challenges of the steep
walls in empty backyard pools beckoned. Eventually, street skating found affordances in everything: ledges, curbs, stairs, handrails—edges and angles of all
kinds. Even with the proliferation of skateparks, pure street skating is still the
true measure of skill and vision. But just as the skateparks have spread, creating
skateable terrain in towns large and small all over the world, the effort against
street skating has evolved as well, creating its own countermovement.

There are several bus bench designs that allow sitting while waiting for mass
transit yet prevent the bench from being used as a bed. Most of these designs
involve armrests or ridges in the seat. In his book *Callous Objects: Designs
Against the Homeless*, Robert Rosenberger writes, "The websites of bench man-
ufacturers rarely advertise the fact that these designs are specifically intended
to discourage sleeping, although on occasion such partitions and armrests are
referred to as 'antiloitering' features."[19] One of the most hostile has to be the
backless, rounded-top bench, a horizontal half-cylinder that allows one to sit,
though not very comfortably, but pretty much prevents lying down. The urban
theorist Mike Davis called these "bum-proof benches" in his book about Los
Angeles, *City of Quartz*.[20]

The manipulation of the perceived affordances of objects and surfaces is
another great example. Chairs and tables offer surfaces that are affordances
for the support of weight. That is, a table affords support. If you have a glass
counter on which you want nothing placed, it should be slanted. If it's flat,
it gives the perception of affording weight placed on top and often ends up
cracked.[21] Handrails around hotel balconies are typically rounded or beveled

Skatestoppers installed on a once thoroughly enjoyed ledge. Photo by
Mark Turnauckas. Reprinted under Creative Commons License.

in such a way as to prevent the setting down of a beverage. This is to keep one
from setting a beer bottle on the rail, then drunkenly or excitedly knocking it
off onto passersby, cars, or just the ground below. These are not design flaws,
they are features.

In the past few decades, architects and urban landscapers have made or ret-
rofitted handrails and ledges to make them unusable for skateboarding. Large
knobs welded onto metal handrails or blocks bolted to ledges keep skateboarders from using these surfaces as props or obstacles for their maneuvers.[22]
These are not mistakes but rather hindrances designed—often clumsily or not
aesthetically—for preventing certain uses.[23] "The architecture of our cities is
a powerful guide to behavior," Alex Andreou writes in the *Guardian*, "both
directly and in its symbolism." He continues, "From ubiquitous protrusions
on window ledges to bus-shelter seats that pivot forward, from water sprin-
klers and loud muzak to hard tubular rests, from metal park benches with solid
dividers to forests of pointed cement bollards under bridges, urban spaces are
aggressively rejecting soft, human bodies."[24] Even if nonverbal, the message is
clear: you are not welcome here.

In the late 1990s, when Nike first attempted to enter the skateboarding market with a line of shoes designed specifically for the sport, their ad campaign
featured unusable athletic spaces like basketball goals permanently blocked
by welded crosses of rebar, football fields with chained-off fifty-yard lines, and

baseball diamonds with obstructed home plates. The ads ran with the slogan, "What if all athletes were treated like skateboarders?"[25] The ads were an interesting take on skateboarding's place in the larger context of sports, but, as discussed last chapter, it would take years and a generational shift before Nike was able to establish itself in the subculture.

"Skateboarding is a rebellion," Miki Vuckovich, a photographer and fixture of the skateboarding industry tells me. "It's a sport, an activity, a lifestyle adopted primarily by adolescents, and as an individualistic activity it gives participants a sense of independence. It sets them apart from others who don't skate, and their own particular approach to skateboarding further differentiates them from their skating peers."[26] When Nike did finally infiltrate skateboarding after the generational changeover, their presence inspired countercampaigns. Consolidated Skateboards, which had launched the "Don't Do It" campaign in 1997, teamed up with Osiris Shoes ten years later to release a Nike "Dunk" knock-off called the "Drunk." The shoe featured a banana replacing the Nike swoosh.[27] Rebels are going to rebel.

Skateboarders are the ultimate nomads of the city.[28] "Skateboarders, particularly adolescents, seek identity," Vuckovich says. "They want to be a part of a clan or tribe that reflects what they're feeling or how they see the world. Most skateboard companies are structurally the same: manager or CEO, a sales/marketing director, a team manager, warehouse and production staff, art director, and skate team. The logos they promote and personalities they collect in their skate team differentiate each brand."[29] The product differentiation in the skateboarding marketplace is infinitesimal: the boards, wheels, trucks, clothing, et cetera vary very little from brand to brand. I mean, how many truly different pieces of plywood with urethane wheels could there be? Most of the products themselves are made by the same few companies. All sales in this industry of image are made on the strength of the brand alone. This strength is usually a product of who is on the team—that is, which professional skateboarders ride for the company. This makes the average skateboard brand a product of the image of the pros on their team and the elusive "coolness" of their name.

An old Nike shoe ad that asks, "What if all athletes were
treated like skateboarders?"

Miki Vuckovich, three wheels out during a backside carve on the east wall of the
Del Mar Skateboard Ranch Keyhole pool circa 1985. Photo by Grant Brittain.
© Grant Brittain. Courtesy of Grant Brittain.

CHAPTER 5

FLEECE THE WORLD

It's all become marketing, and we want to win because we're
lonely and empty and scared and we're led to believe winning will
change all that. But there is no winning.

—Charlie Kaufman

Caught up in the endless traffic and exchange of signs—from billboards,
through television, in newspapers, on film—we construct from this seductive public
rhetoric versions of "reality" to which we give allegiance or in which we place our
faith. . . . But since the flow of signs is constantly changing in the practices which
make up everyday living, since ideas are constantly being modified, disseminated,
reexperienced, re-expressed and transplanted, what is believable changes too.

—Graham Ward

The cyclical nature of the skateboarding industry lends itself to several stand-out eras. The 1970s spawned the Dogtown crew and the Zephyr team, documented on film by Z-Boy Stacy Peralta. The 1980s were all about the Bones Brigade, again led by Peralta and then business partner George Powell under the Powell-Peralta banner. After the reign of the Bones Brigade, ex-pros and other aspiring skateboard entrepreneurs went start-up. During the early 1990s, the skateboarding industry went through one of its down periods and saw a splintering of ownership, a rash of new companies, and an identity crisis of sorts. Large brands like Vision, Santa Cruz, Gull Wing, Independent, and especially Powell-Peralta and their Bones Brigade team dominated the mid-to-late 1980s.[30] When interest in skateboarding waned at the end of that decade, a lot of the older pros started their own companies. One was Steve Rocco. With a six-thousand-dollar cash advance from his credit card, Rocco started what would eventually become one of skateboarding's biggest empires, one that would set trends for many of its followers: World Industries.

The early history of World, and indeed all the new, skater-owned companies that jostled for market share in the early 1990s, is tumultuous. A former skateboard pro and a team manager, Rocco was once told by a company owner that skateboarders couldn't run companies. After getting fired as a team manager, Rocco did just that. He sniped team riders, pirated images from popular culture for board graphics, and concentrated on a subversive, street-smart style that immediately connected with the kids of the time. The intense intricacies of flat-ground freestyle tricks were dead, and the barriers to entry for riding monolithic vert ramps were prohibitive to most. Street skating was anyone's game. Walk out the door, jump on your board, grind a curb, and—congratulations, you're street skating. Focusing on the DIY spirit and the irreverence of youth garnered Rocco unmitigated hate from the established skateboard companies, cease-and-desist orders from copyright holders he appropriated images from, and the attention of millions of faithful followers. "Ads were just so fun," Rocco says of the time. "They were probably the biggest creative outlet I ever had, besides skating."[31] Even with, or perhaps because of, all of this turbulence and chaos, it was also one of the most creative times in skateboarding.

"We would do ads without products," Rocco says, "skateboard graphics with cartoons instead of skulls, and skateboard shapes that didn't look like skateboards. In the wacky world of skateboarding today this sounds like no big deal, but in 1990 people thought we were out of our minds, and at the time, we were. In fact, only in retrospect can we now look back and see the thin line we walked was closer to insanity than the premeditated genius that people often give us credit for."[32] Disruption is another cul-de-sac. Of the Rocco era, Cole Louison writes, "It was just a new extreme, obsessed with its own downfall."[33] Steve Rocco probably wouldn't disagree.

"There are no lines," he told Spike Jonze in a *Dirt Magazine* interview from 1992. "Lines are only in people's heads, and my lines are way different than most people's. I think kids like it when they see somebody pushing it, pushing the limits of what 'adults' would find acceptable. For anything to change, you have to push the old stuff out of the way. Young people do that—things their parents do that are lame, they push out of the way and bring up new stuff."[34]

The products for sale in skateboarding don't differ very much from brand to brand. The subtleties of one board, wheel, or truck are almost indiscernible. A world like that needs a thorough shaking-up once in a while, and a lot of the shaking Rocco did back then is still reverberating today. Most skateboard companies are now run by current and ex-skateboarders, street skating is the most popular aspect of the sport, and, thanks in large part to Rocco's crassly irreverent *Big Brother* magazine, the *Jackass* TV show, and subsequent movies, skateboarding is still disrupting media of all forms.

Conformity is its own reward. Dissent is not. By upending the established order, Rocco brought a lot of grief upon himself. There's the world the way you want it to be, and there's the way that it is. The more dissimilar they are, the more friction you can expect.

FOUNDATION'S EDGE

I don't want to nail things down. I want them to either breathe or explode.
—Laurie Anderson

The name Tod Swank has been a part of skateboarding history since the early 1980s. Besides being a professional skateboarder, he has also been a photographer for *Transworld Skateboarding* magazine and an avid zine-maker. I clearly remember the first time I saw *Swank Zine*. A friend of mine in Alabama got one in the mail, and we all rushed over to his house to see it. They were so rare. To us, it was a photocopied piece of gold. Swank's house sits on a cliff in San Diego that overlooks Mission Valley and the intersection of Interstate 8 and Highway 163. I rode my bike over there through the sculpted neighborhoods and wide streets one Sunday evening in March 2004. "This is what I do every Sunday," he said as we settled in to chat. "I usually do yard work, then come out in the garage, have a beer, watch the sun set."[35]

Originally Swank just wanted to ride for Rocco. "In 1989, I was riding for Skull Skates," Swank explains. "Rocco started World Industries in late '88, and I went to him because I wanted to ride for World Industries, but he wouldn't let me. He said he'd help me start a company, but he wouldn't let me ride for the team. So, I thought about it for a few days and decided to do it even though it wasn't really what I wanted to do."[36] So Swank started Foundation Skateboards—the name comes from the Isaac Asimov science fiction trilogy Foundation—and spent the first two years as part of Rocco's World Industries before venturing out on his own.

"When Rocco started World Industries," Swank continues, "what he really did was liberate skateboarding so that it could move forward. He helped a lot of people start companies, not just me. He lent money and gave advice to a lot of other skateboarders who wanted to start companies. He wanted to see the industry run by skateboarders. The funny thing is, here we are ten years later or whatever, and all of the guys that are running the companies are guys that I skateboarded with or knew before any of us even fathomed that we were going to be working in the industry."[37] In their roles as professional skateboarders and as industry moguls, guys like Swank are culture creators. They have built a subculture that kids identify with very strongly.

"I never intended that," Swank says. "I just wanted to be a part of the skateboarding culture, but at the same time you end up building loyalty and peo-

Tod Swank, pushing, 1987. Photo by Grant Brittain.
© Grant Brittain. Courtesy of Grant Brittain.

ple who are really into it because it's done something different."[38] Foundation is now a part of the Tum Yeto empire that has included such brands as Pig Wheels, Ruckus Trucks, Toy Machine, Deathbox, Hollywood, Dekline, and Zero—all run by skateboarders.

The skateboarding industry provides a microcosmic analog to the rest of the advertising and branding world. Advertising attempts to create a need, an appetite, and then satisfy the appetite with products. Sometimes the products are actually needed, but in many cases they are not.[39] Skateboarding illustrates the manner in which this works with its dearth of differentiating products. If a brand can make it in a world where any product is about the same as any other, and its consumers are staunchly independent practitioners, then it has truly excelled as a brand.

Citing corporate-sponsored educational programs in schools, McDonald's in churches, and phrases owned by corporations as evidence, David Boyle writes, "In this kind of branded reality, people are afraid that their grip on what is authentic is somehow slipping from them."[40] All any of us want is to feel like we belong and to have our experiences mean something.[41] When it comes to skateboarders, that includes knowing you're as much an outsider as a part of something. Vuckovich explains, "Skateboarding, however mainstream it may become, will always give skateboarders a unique perspective on the rest of the world. It's as simple as the difference between crawling on all fours or walking upright. Skateboarders glide around, hopping up and down onto and off of curbs and objects. Benches and handrails are for grinding, not sitting or holding, and staircases are for jumping, not climbing."[42]

"The key to giving an impression of authenticity is rebellion," adds Boyle.[43] When the two biggest markets in America are consumer technology and the escape from consumer technology, resistance is inherent.[44] "The world is a different place on a skateboard," adds Vuckovich. "If you haven't seen it from that perspective, you can't hope to understand a person who has. Or successfully market a product or lifestyle to that person."[45] Since the 1980s, skateboarding has been on the leading edge of many aspects of our mediated lifestyles. Documenting feats and falls, skateboarders were ahead of the endless reality show of social media wherein every moment is captured and shared at its peak.[46] Skateboarding still succeeds in other media as well.

VIRILE MEDIA

One of the few to continue what the architect and author Will Wiles calls "the eschaton of the analogue," *National Geographic*, a print magazine that is over 130 years old, has become hugely successful on social media in the last several years, and in 2017 the company topped 350 million followers across all platforms.[47] In February 2018, editor-in-chief Susan Goldberg told Marketing-Interactive that nothing that *National Geographic* produces is purely for print or for social media. Instead, staff repurpose text, photos, and videos to be shared in different ways to reach different audiences. This kind of strategy requires thinking not only in text and pictures but also about how a story will read in sequence, or out of sequence, both in print and online.[48]

Thrasher is skateboarding's longest-running publication. Founded in 1981 by Fausto Vitello, *Thrasher* is the first magazine created and edited by skateboarders and aimed at recapturing the original spirit of skateboarding—that is, the discovery and manipulation of existing terrain rather than skateparks made for such activities.[49] Its rival *Transworld Skateboarding*, which started just two years after *Thrasher*, presented a cleaner version of skateboarding. It even had nicer paper and full-color photos.[50] Its debut issue featured a response to Craig Stecyk's "Skate and Destroy" article from *Thrasher*'s first issue. Its title: "Skate and Create."[51] "They called us the slick, goody-goody mag, and they were the punk mag," *Transworld* photographer J. Grant Brittain reflects.[52] *Transworld* was eventually purchased by larger and larger media conglomerates that knew nothing about how skateboarding culture worked, while independently owned *Thrasher* grew throughout the 1980s to become the biggest-selling skateboarding magazine during the highly competitive 1990s.[53] I worked in their offices for a brief time back then, in the rugged Hunter's Point neighborhood of San Francisco where they are still located. When I worked there, the *Thrasher* offices had a gravel parking lot, and you couldn't even skate down the street. Still thriving in print, *Thrasher* has always had an edge that other skateboard media lack.

"*Thrasher* is a culture unto itself," *Thrasher*'s longtime advertising director Eben Sterling tells me, "And part of that culture is the print magazine. And I would even go further: within skateboarding as a culture, the magazine has been central . . . *Thrasher* magazine has been essential in building and develop-

Elijah Berle turns a wallride into a grind on the cover of *Thrasher*,
October 2024. Courtesy of High Speed Productions, Inc.

ing and sustaining skateboard culture."[54] How has the longest-running skateboard magazine, one that still strives long past the peaks of most others, been
able to continue successfully?

"One [thing] is advertising," Sterling continues, "The short answer is that
we still have an advertising base. It's kind of an anomaly. That's why the other
ones go out of business. They're not selling magazines, and they don't have advertising."[55] It's a constant feedback loop between the publication and the advertisers. *Thrasher* keeps thriving as a magazine by putting the magazine first,
and advertisers know that they're going to be around to expose their products
to skateboarders.

"Tony [Vitello], the owner, to him, the magazine is number one," Sterling concludes. "He's going to keep putting out the magazine no matter what. If it's not selling enough, if it's not selling enough ads, you still have to put out the magazine. He feels like it is the heart and soul of what *Thrasher* is. If you asked him right now, he would say number one priority: to keep the magazine in print. That's what fuels the entire fucking deal. That's our mystique. That's our myth. That's our story. It's a magazine, and that's why we're still called '*Thrasher* magazine.'"[56]

Like *National Geographic*, aside from staying strong in print, *Thrasher* has adapted and diversified as the media environment has shifted. It has chopped up its content into sharable bits for online and mobile consumption. It has behind-the-sequence features on tricks from the magazine ("Magnified"); the struggles of skateboarders trying to land a trick ("My War"); a weekly news magazine (*Skateline nbd*; shout out to Gary Rogers); a series of tricks all done in one long shot ("Firing Line"); and their ever-popular slam clips of skateboarders eating it hard ("Hall of Meat"). A diversified audience has followed with these separate spreadable clips. After all, people who don't skate still enjoy seeing other people fall down.

I asked the editor-in-chief, Mike Burnett, how intentionally this all came together. "One hundred percent, it's on purpose," Burnett says. "When you come up with an editorial idea, like you try to fit it into one of our departments, and when you create a new property, you try to give it a cool name. . . . You've got to come up with a good name and then be consistent with it. Some of them last longer than others, but the good ones, they've been around quite a while."[57] It's a tactic of branding these properties and making them sharable, mining the content of one medium to spread on another. Not unlike skateboarders mining affordances from the interstices of the city, skate media repurpose what's available.

As noted, the terms *underground* and *mainstream* don't signify the monolithic divides they once did, but there is still a divide. If you look closely enough at a televised "extreme sports" broadcast, you will notice that skateboard companies are not among the many advertisers (such as energy drinks and giant shoe companies). None of the core companies that make skateboards, trucks, wheels, or other skateboarding gear have the funds to buy ad time on television. Likewise, if you pick up a copy of *Thrasher*, you're not likely to see many, if any, ads for energy drinks or shoe companies. This is a problem. It indicates syphoning off resources from a culture without replenishing them. It's not a sustainable situation for anyone involved. Once the money is drained, everyone loses.

Regardless, the spirit of skateboarding will always survive. It has an insatiable drive, and its practitioners will never stop pushing. Magazines and marketing notwithstanding, the built environment remains a veritable playground when seen from a skateboard, an ever-changing obstacle course of possibilities. Ian MacKaye says, "For most people, when they saw a swimming pool, they thought, 'Let's take a swim.' But I thought, 'Let's ride it.' When they saw the curb or a street, they would think about driving on it. I would think about the texture. I slowly developed the ability to look at the world through totally different means."[58] Skateboarding is the built environment dreaming.

The features and forms of our fragmented media culture are designed from the top down by marketing, entertainment, and technology firms, but their meaning is largely determined by our use of them. Remember that the Walkman was originally designed to be used to share music, but once its use by individuals as a personal stereo was noted, Sony retooled the device and the marketing behind it. The Walkman is just one example of how a bottom-up system is also at work in our media culture. The intended design is repurposed by its adopters, then adapted to fit their use. Like desire lines angling away from paved sidewalks, skateboarding is the slang in the pattern language of architecture. Architects call it an "urban pathology."[59] It works constantly against the rhetoric of the built environment. It is where the affordances of design and the desires of humans diverge. Adapting to terrain made for other things is the true test of anyone on a skateboard, a reinterpretation of the nonverbal expressions of the built environment, an appropriation of its many metaphors.

CHAPTER 6

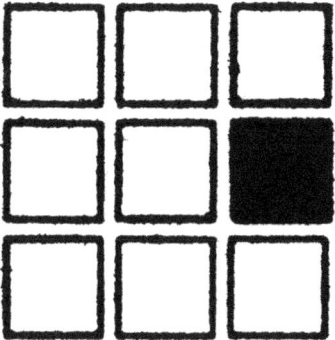

A Message in a Bottleneck

It's such a good metaphor, isn't it? All dreams and memories
are metaphors at last, mere functions of language. Even
without words—they still exist in your head, in one of those
little silent rooms you keep the key to.

—The Addler Clome, in K. W. Jeter's *Noir*

I don't speak. I operate a machine called language.
It creaks and groans, but is mine own.

—Bijaz, in Frank Herbert's *Dune Messiah*

Texts are messages in bottles. The reader is the future of the text.

—Timothy Morton, *The Ecological Thought*

PLANNED ADOLESCENCE

The book in which I think I managed to say most remains *Invisible Cities*,
because I was able to concentrate all my reflections, experiments, and
conjectures on a single symbol; and also because I built up a many-faceted
structure in which each brief text is close to the others in a series that does
not imply logical sequence or a hierarchy, but a network in which one can
follow multiple routes and draw multiple, ramified conclusions.

—Italo Calvino, *Six Memos for the Next Millennium*

The metaphor is perhaps the most fruitful power of man. Its
efficacy verges on magic, and it seems a tool for creation which
God forgot inside one of His creatures when He made him.

—José Ortega y Gasset

The first live concert I saw was KISS. I was eight years old, and there was nothing else like them. The makeup, the mythology, the comics, their characters: the Demon, the Starchild, the Spaceman, the Catman. There was even the 1978 movie *KISS Meets the Phantom of the Park*, a schlocky, science fiction romp in which the KISS superheroes battle with evil robotic versions of themselves. The movie was cartoonishly prescient. It was a lot of stimuli at the time. My dad drove me to see their performance, an hour and a half from Gulfport, Mississippi, to Mobile, Alabama, because we didn't know they would be playing in Biloxi a couple of months later. Such vital information flowed less freely back then.

The physical presence of the band live was terrifying. I looked away when the Demon, Gene Simmons, blew fire. I watched through my fingers when he spit blood during "God of Thunder." I stared, my jaw agape when Spaceman Ace Frehley's guitar smoked during his solo on "Shock Me." Their hold on my young mind went on unrelented. I won a ticket to the December show in Biloxi at a KISS look-alike contest, and, rather than take me to see them again, my dad gave me the ten dollars another ticket would have cost him.

It sounds like the standard fandom of youth, but pop culture moved much more slowly in 1979. The second *Star Wars* movie wouldn't be out for another year. The Atari 2600 gaming system was fun, but it wasn't quite enough to keep me and my friends indoors, off our bikes and skateboards. Things moved slowly enough throughout the 1980s that an album cover or a skateboard graphic would inspire a story that could occupy your mind for months. The wayward space bug on the Journey covers, once captured and then escaped. The monster breaking through the target just a little more on the bottom of every new Rob Roskopp skateboard.[1]

A glimpse at the avatars
of my young mind.
Author's drawing.

I never owned a Journey record or a Roskopp deck, but I would see the images in magazines or at the record store or skate shop, and my mind would wander. I would walk around with these worlds in my head, making connections between things that were never meant to connect, creating narratives, repurposing art and artifacts. That was my young world, and those were the avatars that ruled it, metaphors for adventures untold.

METAPHORS BE WITH YOU

The basic tool for the manipulation of reality is the manipulation of words. If you can control the meaning of words, you can control the people who must use the words.
—Philip K. Dick, "How to Build a Universe That Doesn't Fall Apart Two Days Later"

Technology is not mere tool making and tool use: it is the making of metaphors.
—James Bridle, *New Dark Age*

The biologist Robin Wall Kimmerer writes that when botanists go out in fields and forests looking for flora, they call it a *foray*, and when writers do the same, it should be called a *metaphoray*.[2] Marshall McLuhan and his son Eric open their book *Laws of Media: The New Science* with the claim that each of our technological artifacts is "a kind of word, a metaphor that translates experience from one form to another."[3] That is, each new technological advance transforms us by changing our relationship to our environment, just as a metaphor does with our knowledge. There is a point in both the evolution of technology and the evolution of language where we acknowledge a new term or metaphor and that we are using it, and then there is the space further afield, a "break boundary," where the metaphor obsolesces into general usage.[4] As Eugene Thacker told me, "If metaphors are concepts that we forget are metaphors, then it seems important to remind ourselves of the tropic nature of such central concepts as the genetic 'code.' Not only does this invite us to think otherwise (to think about alternative metaphors), but it is also an invitation to rethink the entire relation between metaphor and materiality itself."[5]

If the explanatory power of the metaphor in use is successful, the metaphor becomes invisible. If a metaphor obsolesces into general usage, it is forgotten as a metaphor.[6] But obsolete metaphors can be brought back to life. Jessica Prinz writes, "The clichés and conventions of everyday talk are re-presented and re-run with slight distortions, twists, and deformations, so that we 'wake up' not only to the life we are living, as John Cage would say, but also to the language that we speak and hear."[7] The beginning, the space between acknowledging a metaphor (and just using it) and how we handle the transition is where the process of technological mediation happens. Something is at stake every time we cross that space.

Meanings are malleable. Words bend and break under the stress of unintended use, abuse, or overuse. Like machine parts pushed past their limits, cogs stripped bare of their teeth, the language we use wears out, weakening the culture that carries it and our knowledge thereof. We use metaphors and metonymies of the machine to explain everything from individual bodies and brains to society and the cosmos.[8] In the Middle Ages, people believed that angels helmed the heavens, manipulating the mechanisms of the sky.[9] Aristotle and Cicero used many anthropomorphic ideas to describe natural occurrences, but the technology of the time, needing constant human intervention, offered little in the way of metaphors for the mind. Since then, the human mind has been compared to the clock, the steam engine, the radio, radar, and the computer.[10] Machines, engines, motors—these are visible, tangible things. As Bettina Knapp writes, "Even more dangerous, perhaps, is the fact that machines increasingly cut people off from nature in general and from their own nature, in particular."[11] The forms of mechanization we most need to watch are the ones we can't see, machinery hidden inside black boxes.

In his book *Mechanization Takes Command: A Contribution to Anonymous History*, originally published in 1948, Sigfried Giedion attempts to elucidate the cause of this splitting from our nature, the break between thought and feeling in modern society. The culprit according to Giedion? Mechanization. He uses a typological approach, moving chronologically through each of his categories: springs (movement), means (hand, key, assembly line), agriculture (gardening, bread-making, meat production), household (chair, table, furniture, feminism, refrigeration), and bath (steam, shower). Giedion also follows how the in-house feminism of Catherine Beecher and "curtailed drudgery and improved organization" led to the further mechanization of the home.[12] He illustrates how early computer developer Charles Babbage informed mechanical engineer Frederick Taylor's time studies, scientific management, and the division of labor of Taylor and Henry Ford, the inventors of modern industrialization as well as corporate structure.

Writing in 1970, Aldous Huxley fretted over machinery's influence on culture as well. While acknowledging the ways in which it freed up time lost to labor, thereby enabling art and culture to flourish, he added, "In 3000 A.D. one will doubtless be able to travel from Kansas City to Peking in a few hours. But if the civilization in these two places is the same, there will be no object in doing so."[13]

RAGE FOR ORDER

"This is real exploring," said Jack. "I bet nobody's been here before."
"We ought to draw a map," said Ralph, "only we don't have any paper."
"We could make scratches on bark," said Simon, "and rub black stuff in."
—from William Golding's *Lord of the Flies*

And the angels, and devils, in machines will be us.
—Ray Bradbury

"More perhaps than machinery," writes the economist John Kenneth Galbraith, "massive and complex business organizations are the tangible manifestation of advanced technology."[14] Institutions, bureaucracies, and organizations like organisms led to the globalization of the machine: processors, keyboards, hard drives, screens, spreadsheets, websites, databases, fiber optic cables, satellites, wireless clouds bulging gray with data.[15] "Intelligent aliens studying us from a distance might well assume that multinational corporations are the most evolved lifeform on our planet," William Gibson adds. "They transcend both the individual and the state."[16] These organizations and their office buildings are tangled with technologies that the architect Le Corbusier collectively called "the apparatus for abolishing space and time."[17] As much as elevators, communication technologies—pneumatic tubes, intercoms, telephones, and computer networks—enabled the function and effectiveness of tall business structures.[18]

"Each person is his own central metaphor," wrote Mary Catherine Bateson.[19] Bateson saw the perceptual processes of the organism as a metaphor for the complexities of the world outside it. With the spread and adoption of personal media and the internet, the network metaphor has crept further and further into our thinking.[20] The center thins out to the edges as the network becomes central, the connections as important as the nodes themselves.[21] In his book *Connected*, Steven Shaviro points out that the network is a fractal. No matter how far down you go, any subsection of the network has the same structure as the network as a whole. He adds, "Neurons connect with each other across synapses in much the same way that Web sites are linked on the World Wide Web."[22] From texts to networks, our minds are permeable.[23] I belabor the point here because we are complicit in the use of these metaphors, and, while they illuminate useful and salient aspects of our world, they simultaneously hide other things from us.[24]

The connections of a network are what gives it its power.[25] In turn, the network gives each node its power. For example, a telephone is only as valuable as its connection to other telephones. Where value normally derives from scarcity, here it comes from abundance. If you own the only telephone or your phone loses service, it's worthless. In addition, each new phone connected to the network adds value to every other phone.[26] But at a certain point, fatigue sets in. Connectivity is great until you are connected to people or companies you would rather avoid. Each new communication channel is eventually overrun by marketers, scammers, and spammers, leveraging the links to sell or shill, forcing us to filter, screen, buffer, or otherwise close ourselves off from the network.[27] There is a threshold, a break boundary, beyond which connectivity becomes a bad thing and the network starts to lose its value.

In their *Laws of Media*, Marshall and Eric McLuhan outlined the ramifications of these media through their *tetrad of media effects*. This states that that every new medium *enhances* something, makes something *obsolete*, *retrieves* a

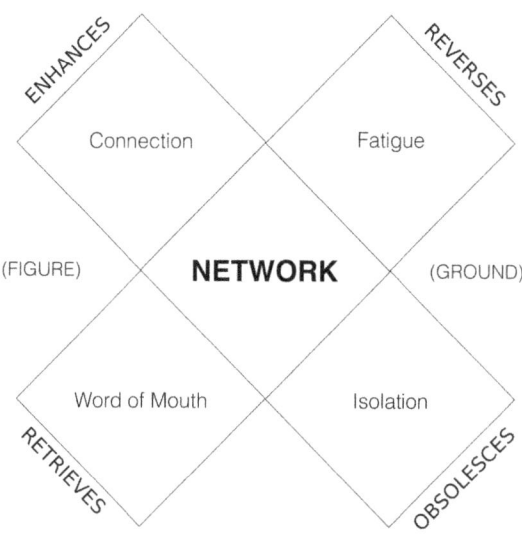

A tetrad of the network. It *enhances* connectivity,
obsolesces isolation, *retrieves* word of mouth, and *reverses*
into fatigue. Author's illustration.

previous something, and *reverses* into something else once pushed past a certain threshold.[28] In the figure above, I have applied this metaphorical framework to the network: it enhances connectivity, obsolesces isolation, retrieves word of mouth, and reverses into fatigue. You can do the same with any new medium, and the McLuhans applied it to many.[29] For example, they asserted that the copy machine enhances the speed of the printing press, obsolesces the mass-produced book, retrieves the oral tradition, and reverses into a lack of readers.[30] When we buy into these infrastructures—networks or otherwise—we buy into their metaphors. Moreover, we buy into the idea that metaphors are an effective way to represent the world.[31]

The designer James Macanufo once said that if paper didn't exist, we would have to invent it. "Digital documents . . . have no edges," writes Lisa Gitelman in her book *Paper Knowledge*.[32] A document in digital space is only metaphorically a document. Every form of media is the same at the digital level.[33] Just as genres of writing emerge from discursive fields according to the shared knowledge of readers, "the ways they have been internalized by members of a shared culture," digital documents are arranged in recognizable forms on the screen, as skeuomorphs, metaphors for familiar objects (e.g., the desktop, the file folder, the page).[34] Like angels in the sky, the underlying mechanisms doing the arranging remain largely hidden from us as users. If paper didn't exist, we would have to invent it. Would anyone say the same for the screen?

Rita Raley describes screen-based, "born-digital" works as unstable, "not texts but text effects."[35] She moves away from viewing the digital document and other such contrivances as metaphors and toward employing what the media theorist Alexander Galloway calls "the interface effect."[36] Galloway's view casts the old argument of interfaces becoming transparent and "getting out of the way" in a bright and harsh new light, and he writes that their "operability engenders inoperability."[37] Once our devices obsolesce into general use, "those transparent devices that achieve more the less they do," as Galloway puts it, they escape everyday criticism.[38] The interface stuff hides in those edges that aren't really there. The words I write now float and flicker on a page on a screen in a conceptual space I barely understand, black-boxed beyond my fingertips. Aside from the shared metaphors of these encoded experiences, the underlying mechanisms doing the arranging remain largely hidden from us as users.[39] We are the nodes in this invisible network, individual monads reflecting the nodes we are connected to.

MIRRORING MINDS

This proud picture of human grandeur is unfortunately an illusion
and is counterbalanced by a reality that is very different.
—C. G. Jung, *The Undiscovered Self*

Duncan Watts and Steven Strogatz were the first scientists to mathematically describe what we now know as *small-world networks*, the invisible webs that connect certain categories of things by six degrees of separation. The neural network of a worm, the power grid of the Western United States, and the collaborations of famous film actors all exhibit these connective characteristics.[40] Watts and Strogatz made this discovery while investigating how massive collections of crickets synchronize their chirps.[41]

Similarly, when sparrow-weavers sing duets, their brains connect and synchronize to the point of becoming one big bird brain. Researchers from the Max Planck Institute for Ornithology in Germany found that the brain activity of one singing bird changes and synchronizes with its partner when the partner begins to sing.[42] Through a similar neurological study of zebra finches, they also found that the birds can anticipate their mate's calls and speed up their communication.[43]

This kind of cognitive entanglement is not unique to crickets and song-birds. Scientists have found a similar phenomenon in human brains. So-called *mirror neurons* were first discovered in Vittorio Gallese's Italian laboratory. The researchers were recording patterns from a cluster of neurons in mon-key brains. During a break, one of the monkeys observed another peeling a banana, and the part of his brain that would be active if he were peeling it himself lit up.[44] The odd neural firing had the neurobiologists checking and rechecking their equipment. The discovery of mirror neurons has been touted as one of the most important findings of the past two decades, right up there with—and possibly closely related to—language acquisition and imitation.[45] Mirror neurons may also be responsible for our ability to empathize with each other and other living things.

Even at their most simplified, the linking of observing actions and brain stimulation, mirror neurons help explain why our media is so effective in occu-pying our attention, and how watching movies, TV shows, and sporting events, as well as reading novels and playing video games—vicarious experiences of all kinds—make us feel as if we are doing the things we are watching. "Each monad mirrors the whole world," writes Michael Heim in his book *The Meta-physics of Virtual Reality*.[46] It is another compelling metaphor regardless of how well it reflects reality.

MYSTERY LOVES COMPANY

The map had been the first form of misdirection, for what was a map but a way of emphasizing some things and making other things invisible?
—Ghost Bird, in Jeff VanderMeer's *Annihilation*

What is the importance of placing a memory? he said. Why spend that much time trying to find the exact geographic and temporal latitudes and longitudes of the things we remember, when what's urgent about a memory is its essence?
—from Steve Erickson's *Days Between Stations*

In his 1981 short story "Johnny Mnemonic," Gibson writes, "We are an infor-mation economy. They teach you that in school. What they don't tell you is that it's impossible to move, to live, to operate at any level without leaving traces, bits, seemingly meaningless fragments of personal information. Fragments that can be retrieved, amplified."[47] Sometimes the fragments just sync up with no

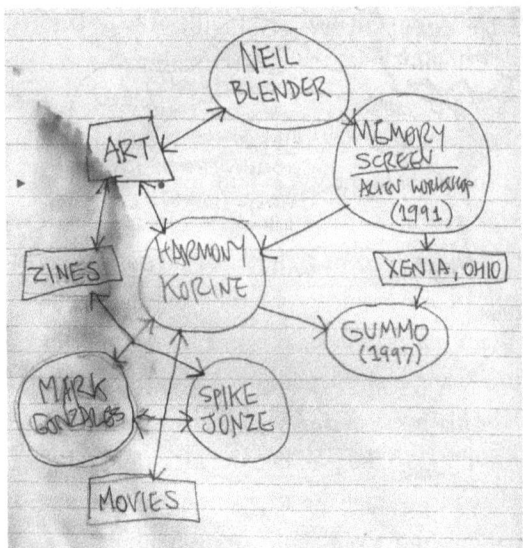

A flowchart of associative trails from one of my notebooks.

meaning outside of the alignment itself. Sometimes it's a winding wandering with no logic whatsoever, nodes emerging in an indecipherable network. Other times the connections themselves are the important part. Sometimes it's the difference between searching and finding.

Occasionally I get a book in the mail or from the library, and I can't remember how I found out about it. In late 2000, during an especially impoverished period of my adult life, I was going to the Seattle Public Library almost every day. I was reading bits and pieces of so many books. I remember digging deeper into the work of Marshall McLuhan and Walter Benjamin, finding more Rebecca Solnit and Guy Debord, discovering Paul Virilio and Jean Baudrillard. I remember the row of volumes I had lined up against the wall in an almost completely unfurnished apartment, their spines and call numbers pointed at the ceiling. Due dates, new discoveries, and new arrivals kept the books rotating, and at some point I started having a difficult time keeping up with where I had read what. So I started keeping a research journal. While those journals—I've been keeping them ever since—help me remember things I have read after I've read them, they are less helpful in tracing my path to the things I read in the first place.

I have tried to keep up with the latter paths with flowcharts like the one in the figure above. This is not a new idea. Early versions include Paul Otlet's 1934

Traité de Documentation and Vannevar Bush's memex from the 1940s. The memex—"memex" is a portmanteau of *mem*ory and *ex*pansion—was a dream machine for navigating and researching with the vast stores of information of the time using cameras, microfilm, and print—an annotated analog hypertext system. In his 1945 article "As We May Think," Bush wrote, "Wholly new forms of encyclopedias will appear, ready-made with a mesh of associative trails running through them, ready to be dropped into the memex and there amplified."[48] To retrofit an analogy, in Bush's analog assemblage, McLuhan's *figure* and *ground* are reunited and ready at hand.

The problem with the way that Bush proposes such "associative trails" is that they are of little use to anyone aside from the researcher who has trod them. Do you find the flowchart in the figure above useful without further context? As Donald Norman puts it, "Following the trails of other researchers sounds like a wonderful idea, but I am not convinced it has much value. Would it truly simplify our work, or would all the false trails and restarts simply complicate our lives? How would we know which paths would be valuable for our purposes? . . . If you, the reader, were to follow these trails, you might not be enlightened."[49]

In a more recent example from *Passion of the Weiss*, author Will Hagle follows an algorithm through associative trails on the music-discovery app Shazam. If you hear a song in public, and it sounds familiar or just enjoyable but you can't identify it, the app will listen and look it up for you. "Except I don't usually ever listen to the songs I Shazam," Hagle writes. "My phone just automatically saves a big list of artist names and track titles. Finding out that info is gratifying in the moment, but it's rare that it goes beyond that."[50] It's a network of songs in which the nodes fade in importance, memories without significance. Trees without forests, all limbs and leaves, no roots. Hagle continues: "I have vague memories attached to the songs, because of where I was when I Shazammed them. None of these moments have interesting stories associated with them. I heard the songs on the radio or at a coffee shop or bar. But just seeing the song titles and the accompanying dates gets my memory gears turning. The details are unimportant, except to emphasize that everyone who uses Shazam probably experiences a similar phenomenon. The less memorable the song, the less memorable the memory attached to the song."[51] Here the associative trails of the songs are of less significance than their individual memories.

In contrast, William Gibson is often asked how he predicts the future so well. "I think I'm drifting around looking for nodal points," he tells the theorist Kodwo Eshun. "I put myself in the way of huge floes of information, none of it terribly interesting or important in itself, and look for the points at which I somehow feel change is about to emerge. . . . Nodal points are what make foolish people think I'm prescient. In retrospect, I think I was looking for a metaphor for whatever it is that I do that people mistake for a predictive capacity."[52] Gibson wrote the metaphor straight into his 1996 novel *Idoru*, in the character Laney: "He wondered vaguely if there might be a larger system, a field of greater perspective. . . its own nodal points, infofaults that might be followed down to some other kind of truth, another mode of knowing, deep within gray shoals of information."[53]

Similarly, the information architect Peter Merholz argues that the nodes and connections of networks can reveal the digital equivalent of desire lines. Merholz notes, "A smart landscape designer will let wanderers create paths through use, and then pave the emerging walkways, ensuring optimal utility."[54] As we'll see in chapter 7, this practice is generally unfeasible due to time and budgetary restraints common to building and landscaping projects. He continues, "Once you have a preliminary system in place, you can use the most common tags to develop a controlled vocabulary that truly speaks the users' language."[55] Ironically, this insight was born from Merholz's inability to find something in a database that he knew to be there.

Maybe the mystery is better. I have so many precious books and movies and songs and ideas that lack provenance or pedigree. I might wonder how I found them, but does it matter? Whether by an alignment of fragments or a meandering path, neither changes the result. In between the persistent moments burned in our young brains, and the gaps, blind spots in our histories, myths form from memories. These moments are nodes in a network of novelty. They form a lattice of coincidence. They hint at patterns that need parsing. Meaning lurks around the edges of things, if only we are paying attention.

UNDERMINING MEDIA

The culture industry makes people believe that they participate in culture.
—Jørgen Nash, Jens Jørgen Thorsen, and Dieter Kunzelmann, "Slogans"

Our media are our metaphors. Our metaphors create the content of our culture.
—Neil Postman, *Amusing Ourselves to Death*

Reclaiming the dominant metaphors of a given time is an act of magical resistance. Feigning immunity isn't a solution but does provide a deeper diagnosis of the problem.[56] Appropriating language, mining affordances, and misusing technology and other cultural artifacts can create the space for resistance not only to exist but also to thrive.[57] Aggressively defying the metaphors of control, the anarchist poet Hakim Bey termed the extreme version of these appropriations "poetic terrorism." He wrote, "The audience reaction or aesthetic-shock produced by [poetic terrorism] ought to be at least as strong as the emotion of terror—powerful disgust, sexual arousal, superstitious awe, sudden intuitive breakthrough, dada-esque angst—no matter whether the [poetic terrorism] is aimed at one person or many, no matter whether it is "signed" or anonymous, if it does not change someone's life (aside from the artist) it fails."[58]

Echoing Bey, artist Konrad Becker suggests that dominant metaphors are in place to maintain control, writing, "The development in electronic communication and digital media allows for a global telepresence of values and behavioral norms and provides increasing possibilities of controlling public opinion by accelerating the flow of persuasive communication. Information is increasingly indistinguishable from propaganda, defined as 'the manipulation of symbols as a means of influencing attitudes.' Whoever controls the metaphors controls thought."[59]

In a much broader sense, so-called "culture jamming" is any attempt to reclaim the dominant metaphors from the media. Gareth Branwyn writes, "In our wired age, the media has become a great amplifier for acts of poetic terrorism and culture jamming. A well-crafted media hoax or report of a prank uploaded to the Internet can quickly gain a life of its own."[60] Culture jammers, using tactics as simple as modifying phrases on billboards and as extensive as impersonating leaders of industry on major media outlets, expose the ways in which corporate and political interests manipulate the masses via the media.[61] In the spirit of the Situationists, culture jammers employ any creative crime that can disrupt the dominant narrative of the spectacle and devalue its currency.[62]

What if there are no longer any dominant metaphors to subvert? What if the center of our culture has already been devalued, splintered into subgroups, and pushed out to the edges? Chris Riley writes, "When mass media began to break down with cable television and ultimately social media, individuals discovered an easy default: Find the narrative they preferred, one more resonant with who they actually were."[63] Ours has been described as an attention economy for decades now, but it has since split and splintered.[64] "There is nothing like Debord's grand spectacle," adds Steven Shaviro in his book *Connected*, "no totalizing system of false representations that would masquerade as actual life. Instead, we have a plethora of tiny spectacles, each of which calls explicit attention to its own status of merely being a spectacle."[65] When we all exist in our own individual media silos, how can anything coalesce into a cohesive phenomenon that defines an era or generation? Shaviro continues: "Each spectacle is a monad, entirely self-contained and self-enclosed, yet connected over the network to all the rest."[66] For better or worse, the network is one of the defining metaphors of twenty-first-century culture so far, the connections that bind it together and the artifacts and ideas that spread.

"A metaphor is always a framework for thinking, using knowledge of this to think about that," Bateson once said.[67] The word *metaphor* means "carrying over," and that's just what their meanings do. Machines, mirror neurons, nodes, and networks: eventually, we forget all of these are metaphors. "That is the real danger," Robert Swigart writes, "unless we pause from time to time to consider how these metaphors work to create boundaries. . . they will control us without our knowledge."[68] As long as we are paying attention, though, we can always defy them.

PART III
CROSSING

The Human Patina

Write to the nth power, the n − 1 power, write with slogans: Make
rhizomes, not roots, never plant! Don't sow, grow offshoots! Don't be one
or multiple, be multiplicities! Run lines, never plot a point! Speed turns
the point into a line! Be quick, even when standing still! Line of chance,
line of hips, line of flight.

—Gilles Deleuze and Félix Guattari, *A Thousand Plateaus*

Thus we cover the universe with drawings we have lived. These
drawings need not be exact. They need only to be tonalized on the mode
of our inner space. . . . Space calls for action, and before action,
the imagination is at work. It mows and ploughs.

—Gaston Bachelard, *The Poetics of Space*

Every living thing follows along a set path.

—Donnie Darko

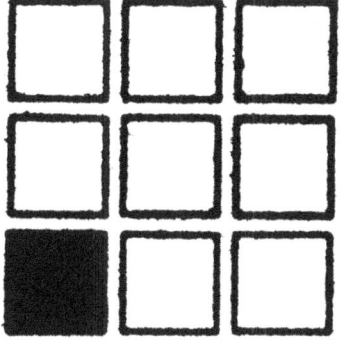

Disguise the Limit

I will not follow where the path may lead,
but I will go where there is no path,
and I will leave a trail.
—Muriel Strode, *My Little Book of Prayer*

Every time I write about my heart, I write about walking.
Every time I write about walking, I write about my heart.
—David Toms, *Pacemaker*

The ground on which you walk is the
tongue with which I talk.
—Saul Williams, "Elohim (1972)"

THE WHEELED AND THE WINGED

Form is your footprints in the sand when you look back.
—Ronald Sukenick

The line is like a rough sketch of a calming and
stabilizing center in the heart of chaos.
—Gerald Raunig, *Dividuum*

A wing is a bridge. Flight is a ride on that bridge from takeoff to landing. As biological forms, dinosaurs became bipedal, balancing their large bodies on two legs, eventually sprouting wings and counterbalancing tails. The same process or set of processes between biology and environment morphed their bodies into bridges and thereby enabled flight. Other flying or flocking things—drones, swarms, crowds, organizations—are insect analogs.[1] Where insects seem to add extra limbs and wings as needed, dinosaurs developed balance along an axis. Theirs was a linear transition from quadruped to biped to winged flight.[2] It was an appropriation of existing limbs, the farming of a new form.[3]

"First, activities of the forelimbs and tail became separated from those of the hindlimb, pelvis and torso," Pat Shipman explains in her book, *Taking Wing: Archaeopteryx and the Evolution of Bird Flight.* This freed up the forelimbs for other purposes, while the rear limbs grew accustomed to holding their own. She continues, "Thus, logic, anatomy, and paleontology all support the same deduced sequence of evolutionary changes: bipedalism first; wings second; tail third."[4] Not all wings are created equally. Not all wings are made for flying. Some enable related abilities such as walking on water.[5] Some are made to regulate body temperature.[6] Nonetheless, the shift in balance (from four legs to two) and the transfer of that balance from one plane to another (from linear—front-to-back—to lateral) informs the evolution of our own flying machines.[7]

Among human-powered tools, the bicycle is one of the most efficient, using the structure of the human body and its musculature in a near-optimal manner. Many early tools used mostly the arms and back, leaving the larger muscles—the legs—to stabilize or anchor the action. Digging, sawing, hoeing, shoveling, chopping, and rowing all leave the legs for standing. It wasn't until treadmills were used to power pulleys and mills that human legs got their due. Inventor Baron Karl von Drais, who appropriated the idea either from a wheelbarrow or ice skates, is widely considered the inventor of the bicycle. In the early 1800s, he put the wheels one in front of the other and demonstrated the ability to effectively balance along their unified line.[8]

It's no wonder that an initial unassisted bike ride often feels like flying. Pedaling that two-wheeled bridge of balance is like taking off on wings of one's own.[9] In more sober tones, Marshall McLuhan aligned the two activities as well, writing, "It was the tandem alignment of wheels that created the velocipede and then the bicycle, for with the acceleration of wheel by linkage to the visual principle of mobile lineality, the wheel acquired a new degree of intensity. The bicycle lifted the wheel onto the plane of aerodynamic balance, and not too indirectly created the airplane. It was no accident that the Wright brothers were bicycle mechanics, or that early planes seemed in some ways like bicycles."[10]

Most early pilots started off riding bicycles. Bicycle racer Glenn Curtiss barely beat his rival Orville Wright in getting the first U.S. pilot's license. The first two German pilot's licenses went to bicycle racers August Euler and Hans Grade, and in France the bicycle-racing Farman brothers also took to the skies and the manufacturing of flying machines. In Italy, Alessandro Anzani, bicycle sprint champion, became an early champion of aircraft manufacturing. These are not coincidences. The same skills required for banking and balancing a bicycle transfer to negotiating the air.[11] "For the leading cyclists of the early 20th century," wrote McMahon and Graham, "flying was a logical extension of cycling."[12]

Though they contributed much to aeronautics, the Wright brothers' crowning contribution to flying came from realizing the need for lateral control of the vessel early in their design.[13] The banking and balancing of a bicycle take on new complexity once the wheels leave the ground, but the linear alignment of those wheels is the same as the axis along which an airplane is guided.[14] So it makes sense that one kind of balance begot another. As the bipedal dinosaur became the flying dinosaur and the bird, our bicycles became the airplane. Their line of chance and their line of hips became their line of flight. Their lines, once landlocked, leapt singing into the skies.

Recall that the body is the original site of all of our media.[15] It is up to the mind to make meaning of language and signs. So, when the gestural cues of buttons and levers become the symbolic cues of images on touchscreens, interfaces shift from the muscular to the mental.[16] The bicycle is also an open system. Unlike the black box of the Walkman or smartphone, it invites tinkering and wear.[17] Also unlike the flat rectangle of the smartphone, its affordances and limitations are visible.[18] Its interface is attuned to the human body, adjustable to multiple uses and terrains and every rider's size and skills, and the labor involved in operating it is mostly muscular. The bicycle is not a spacecraft or a smartphone, but its assemblage of grips and levers and pedals is a nice example of the affordances of a body-based, human-centric, analog interface or simple machine. Its constraints are also its strengths. We love bicycles because we know what to expect from them.

A CHORUS OF VOICES

I'm not interested in legacy in terms of my reputation.
I am, however, interested in leaving a trail.
—Ian MacKaye

"Our world is made up of lines," writes the theorist Anne Seymour, "from comet tails to DNA." Our existences and experiences are summed up by the lines we make, the lines we follow, and the lines we cross. Seymour continues, "Everything is connected. Everything is sequential. Everything that moves, from a snail to a lava flow, leaves a line, a trace of its passing. A line can be fate, a commitment, a fact, a relationship, a place. Some lines are well-trodden paths, some intersect, some pass at a distance, some return to their origins. We all walk the line. We have an end and a beginning, which is joined to a much larger invisible line in the past and in the future."[19]

Lines compress the micro and the macro, from the lines on the ground to the lines on a map, from lines in a poem to lines in a conversation. They illustrate the movement of the individual and the movement of people as a whole. "The life of a person is the sum of his tracks," writes cultural anthropologist Roy Wagner, "the total inscription of his movements, something that can be traced out along the ground."[20] As long as we have been speaking and writing, we have been drawing lines. In his book *Lines: A Brief History*, Tim Ingold stresses that our lines in the world are multidimensional, writing that to move along a path "is to thread one's way *through* the world rather than routing from point to point *across* its surface."[21] In contrast to mapping a route, traversing it on foot is to be a part of the world.

Rebecca Solnit has done perhaps the most thorough job of exploring the history and philosophy of walking and thinking. In her book *Wanderlust: A*

"Untitled (line) 1973," by Liliana Porter.
© Liliana Porter. Courtesy of Liliana Porter.

History of Walking she writes, "Walking allows us to be in our bodies and in the world without being made busy by them. . . . Walking is a mode of making the world as well as being in it."[22] The natural rhythms of walking, breathing, and thinking synchronize as we make our unhurried way through the world. We illustrate this thought in lines both walked and written, lines that Michel de Certeau equates, writing, "The act of walking is to the urban system what the speech act is to language or to the statements uttered."[23] He adds, "Walking affirms, suspects, tries out, transgresses, respects, etc., the trajectories it 'speaks.' All the modalities sing a part in this chorus, changing from step to step, stepping in through proportions, sequences, and intensities which vary according to the time, the path taken and the walker."[24] In other words, we walk the talk. In spite of urban constructs designed to keep us in line, we still meander, stroll, roam, rush, and return according to our own spirits and schedules. The city is a language, and to walk is to speak.[25] "There is a rhetoric of walking," continues de Certeau. "The art of 'turning' a phrase finds an equivalent in an art of composing a path."[26] If maps are voyeuristic, then walking puts one in the middle of the action, making marks, leaving traces, living the lines.[27] Desire lines are not only about the relationship between walking, body, world, and each other but also between walking and design. Maps are metaphors and often represent a bit of both, as well as the relationships between body and world.[28]

William Gibson's novel *All Tomorrow's Parties* has a character, Fontaine, who is a curio dealer. "Fontaine would make up his own story, read function in the shape of something, read use in the way it was worn down. . . Everything . . . had a story. Each object, each fragment comprising the built world. A chorus of voices, the past alive in everything, that sea upon which the present tossed and rode."[29] Indeed, our cities are alive with the stories of their inhabitants. Italo Calvino wrote in his novel *Invisible Cities*, "The city . . . does not tell its past, but contains it like the lines of a hand, written in the corners of the streets, the gratings of the windows, the banisters of the steps, the antennae of the lightning rods, the poles of the flags, every segment marked in turn with scratches, indentations, scrolls."[30] Like lyrics remembered when the song is forgotten, the paths we follow during our daily journeys are more memorable and more meaningful to us than the places we have been and the places we go.[31] It is hard to unsee a pattern.

A FOOTPATH NAMED DESIRE

We are here to inscribe ourselves on the universe, and it is not inappropriate
to remind ourselves of this when blank slates are given us.
—from Kim Stanley Robinson's *2312*

Roads no longer merely lead to places, they are places.
—John Brickerson Jackson, *A Sense of Place, a Sense of Time*

Campus sidewalks meander between places of interest, connecting buildings
and parking lots in a maze of concrete stripes. Often where their right angles
turn near grassy areas, arrowed toward another building or parking lot, there are
paths leading off diagonally. These forking paths are called *desire lines*, so named
because they show where people would rather walk.[32] The folklore of desire lines
says that good engineers—or lazy ones, depending on who tells the story—put
sidewalks in last so as to follow the paths in use and avoid wear on the grass.[33]
The story is suspect, not only because it makes too much sense but also because it
runs counter to the bureaucratic grain. Construction contracts and fiscal funding don't last long enough for engineers to wait and see where people want to
walk. The time constraints wouldn't allow much in the way of paths.[34] However,
there are cases of rogue paths being "legitimized" with pavement after the ones
in place proved insufficient.[35] Impressions of desire take time.

Desire lines are the earth's memory, physical graffiti drawn with feet and
tires.[36] Some are deviations from the built environment, and some precede the
building. Some show what users of a system would rather do, and some show
which procedures they perform the most. That is, sometimes the "desire" refers
to a transgression of the system, and sometimes it refers to normal but frequent
use. Urban planner Kevin Lynch wrote, "Paths can link places of contrasting
episodic quality, as in the stroll garden. On more formal occasions, as on an
educational trip, the journey in space may deliberately symbolize a journey
in time."[37] The "well-worn ribbons of earth" walkers leave behind symbolize
"yearning," in the view of traffic engineer John La Plante.[38] The earth remembers the walks written out by soles.

It is not only about the relationship between walking, body, world, and each
other but also between walking and design.[39] If the built environment is the
external manifestation of our thinking brains, if the urban system is our cognitive system extended in urban form, then desire lines are its deviant thoughts.[40]
If pavement formalizes the "historical layers of a city," then desire lines are its

marginalia.[41] "The design of an office space, or a university, or a prison, or a factory, is likely to embody the power relations that structure its activity," writes the historian Stuart Ewen. "It will tell us, silently, who is powerful, who is not; who watches, who is watched; who dictates, who takes dictation."[42] Silent acts of resistance, desire lines are the transcripts of everyday defiance.[43] The architect Christopher Alexander wrote, "When large buildings are built, people and departments are treated like objects, and bundles of them are allocated to holds inside a cargo ship. Under such conditions, how is it possible to feel any sense of ownership and responsibility? How is it possible to care for one's environment, and make plans to change it?"[44]

Desire lines illustrate the tension between the built environment and our relationship to it. Desire lines are where the system—the system of people in conjunction with their built environment—asserts itself. Designing and laying sidewalks are predictive behaviors. They anticipate the future. Desire lines are signs of past and present use. They are the site of discrepancies between designers' intentions and our actual behaviors, and they are the best guide for path-building.[45] In a passage worth quoting at length, Kevin Lynch and Gary Hack describe lines flowing through the city:

> The character of the line should depend on the speed with which it will be traversed. A footpath responds nervously to minor changes in terrain; a highway takes a sweeping line. Pedestrian motion, like the flow of water, has an apparent fluid momentum. It follows the lines of least resistance, shortening distances by cutoffs. The flow may be smooth or turbulent, purposeful or meandering. It can be deflected, or encouraged by visual attractions, by levels, openings, and the character of the floor. A walk may be arcaded, heated, cooled, or its floors warmed to melt the snow. It can be provided with benches, plants, kiosks, cafes, display cases, or information devices. As a fine highway expresses the nature of vehicular movement, so a good walk system reflects the pleasures and characteristics of motion on foot.[46]

As Lynch and Hack describe it, design should be tempered by the act of walking, but the system itself determines the tone of the walk.

The most famous philosophers of the urban walk, Guy Debord and the Situationists, walked their own talk, drifting through cities with remixed maps and minds set on getting lost. It was their way—the method called *psychogeog-*

raphy, the drifting called the *dérive*—of experiencing the urban environment deliberately.[47] "The *dérive*," wrote Debord in 1958, "entails playful-constructive behavior and awareness of psychogeographical effects; which completely distinguishes it from the classic notions of journey or stroll."[48] That is, these are walks of resistance, against the schedules of sidewalks and the clockworks of concrete. They embrace the ephemeral nature of their environment.[49] And they are supposed to be fun![50] "The point was to encounter the unknown as a facet of the known," writes Greil Marcus, "attending to a map of your own thoughts, the physical town replaced by an imaginary city."[51] Forbidden places were especially enticing for their less-trafficked, forgotten nature and their fulfilment of mischievous desires.[52] As if describing Alexander's traffic-flow diagram below (see page 129), the Italian architect Francesco Careri describes one of Debord's maps:

> The city is nude, stripped by the *dérive*, and its garments float out of context. The disoriented quarters are continents set adrift in a liquid space, *passional* terrains that wander, attracting or repulsing one another due to continuous production of disorienting affective tensions. The definition of the parts, the distances between the plates and the thickness of the vectors, are the result of experienced states of mind.[53]

The *dérive* divines the multiple, invisible cities housed in the consensus city its citizens agree to see. Drifting through the city streets "in search of signs of attraction or repulsion" is perhaps the simplest act in the arsenal of Debord's persistent pursuit of authentic experience: a small, subtle gesture aimed at clawing at the cage of the society of the spectacle until it turns inside out.[54]

POWER LINES

To know someone was to know what to expect. People were as
much trajectories as they were face, form, or voice.
—Tom, in R. Scott Bakker's *Neuropath*

A path was a path was a path was a path. A path was a
person and a path was a mind, walk, chop, walk, chop.
—Patricia Lockwood, *No One Is Talking About This*

Desire lines remind us that aspects of our lives only matter because a certain number of us have decided that they do. Some call it social construction, and, though it is often criticized as uselessly postmodern, the concept is testable. Go to your local coffee shop or restaurant and try to walk behind the counter. You will be swiftly ushered back to the other side of the counter if not out of the establishment.[55] Whether or not there is an actual physical barrier in place, there is an accepted area for the employees and one for the patrons—that is social construction. We are complicit in the power possessed by these boundaries.[56] As a society or culture we tend to agree on a great many of these constructions. We decide what matters.

In addition to the destruction of landscaping and pockets of nature near buildings and paved paths, part of the reason desire lines are such a nuisance for engineers is their permanence. Once a path is in place, it is difficult to deter people from using it. "Most plants die if stepped on more than a few times," write Bruce Hampton and David Cole, "and unstable soils start eroding even with light trampling. Once these processes begin, the impact quickly accelerates."[57] In a pathetic pursuit of order, the powers in place want to keep us in line, implementing technology that makes decisions for us. Solnit writes, "Just as language limits what can be said, architecture limits where one can walk."[58] Paved sidewalks are predictions, attempts at restricting the walks of the future (*top-down*). Trails are of the past, worn by many previous walks (*bottom-up*).[59] Often we aren't left a choice as to what is easier, more convenient, or more fun, much less what is more acceptable. Often the technology in place makes only one path available—a sidewalk in the current example. But, Solnit continues, "the walker invents other ways to go."[60] GeorgieR, an admin for an online group dedicated to the study and catalog of desire lines, writes, "The key to the desire path is not just that it's a path which one person or a group has made, but that it's done against the will of some authority which would have us go another, rather less convenient, way."[61] Regardless of the power of pavement, desire lines illustrate our endless ability to stray anyway, to make paths of our own where there were none before.

The archaeologist Christopher Tilley writes, "Paths are . . . fundamentally to do with establishing and maintaining social linkages and relations between individuals, groups and political units. Social paths and the paths followed through the forest may become overgrown through lack of use in any particular (physical or social) direction."[62]

"An obvious path soon attracts others," write Hampton and Cole.[63] We make sense of the world based on the evidence presented to our senses that we find consistent with our past experience. Wayfinding in any environ-

Traffic-flow diagram
from Christopher
Alexander's *Notes on
the Synthesis of Form*
(1964). Copyright © 1964 by the
President and Fellows of Harvard College.
Copyright © renewed 1992 by Christopher
Alexander. Used by permission.
All rights reserved.

ment—natural or urban—is based on frames of reference and memories
but also what route looks the safest. Anticipating de Certeau's walking-as-
writing, urban planner Kevin Lynch called a city with a recognizable, coherent
pattern a "legible city." He wrote, "Just as this printed page, if it is legible, can
be visually grasped as a related pattern of recognizable symbols, so a legible city
would be one whose districts or landmarks or pathways are easily identifiable
and are easily grouped into an over-all pattern."[64] In a legible city, "an obvious
path" attracts more feet because it makes sense. It follows the grammar of the
daily journey. A preexisting desire line indicates to walkers a way to go, possibly
a better, safer, or shorter one. Like highlighted lines in a well-worn book, paths
reveal knowledge that previous visitors found useful. This is the basic assump-
tion underlying the networked knowledge of the Web's interconnected hyper-
links and the citations, notes, and references in a book derived from the work
of others like this one.[65] Not every sign of wear or use is a desire line, but they
do easily lend themselves to metaphor.[66] Wayfinding through an environment
of words or woods isn't instinctual. It happens based on such environmental
clues.[67] Walkers are more likely to walk where others have already, and that is
not always where planners would have them walk.

In the simple traffic-flow diagram above, the thickness of the lines illustrates the amount of traffic, and the arrows designate the direction of the flow.
"Clearly a thick arrow requires a wide street," wrote Christopher Alexander, "so

that the overall pattern called for emerges directly from the diagram."[68] Piles of data like this are used to design or redesign urban transit systems.[69] The thick arrows here represent what Mark Rose calls "more desirable lines" in that they illustrate the path people would rather take given the choice among all possible paths.[70] Designers use such information in attempts to accommodate the needs of the users of mass transit. Where desire lines are often a matter of avoidance, leading around obstacles or across expanses toward a shorter path, here they are a matter of affordance.[71]

In an influential example, the 1955 Chicago Area Transportation Study (CATS) planners defined a desire line as "the shortest line between origin and destination, [which] expresses the way a person would like to go, if such a way were available."[72] To them, these lines are less about desire and more about measurable behavior.[73] Providing paths and transit in line with city travelers' wants and needs is better for all concerned.[74] Chicago's layout stands at sharp right angles to the improvised paths of its citizens.[75]

One hundred years earlier in Chicago, a mid-nineteenth century attempt at a public square as a center of "civic engagement" among the tallest buildings downtown ended in messy trails. "Muddy and unkempt, it was a shortcut site in contrast to the grid in whose hypothetical center it was located," writes historian Peter Bacon Hales. "Its failure was its success; offering an alternative to the regulated patterns of movement within the built-up blocks surrounding it, the open square increased the efficiency of those who moved through it, while losing its place as a greensward."[76] In 1851 the site was chosen for a planned new government building, which by 1871 took up the whole block. Putting an entire building in the way might seem extreme, but keeping errant walkers controlled not only prevents further wear where planners would rather there be none but also limits other kinds of damage. "Broken windows theory" states that urban litter, graffiti, and broken windows are the slippery slope down which a community slides into more serious disorder and crime.[77] If the neglected aesthetic features of an area indicate one set of bad behavior, then worse is sure to follow. Vandalism ignored is the gateway to more serious offenses. Though the theory has been critiqued as too narrow in scope, it isn't difficult to see its logic where desire lines are concerned.[78]

Before they were the slang of the pattern language, a blight on the urban planner's finished project, desire lines prefigured roads and maps. Before the first roads were paved, they were dirt paths worn by hooves and wooden wheels.[79] Before that, they were trade routes trampled by footfall.[80] And before that, they were simply paths indicating a desire to find a way. Writer Peter Turchi explains, "Tens of thousands of years ago, before the first trails were etched into mud with the point of a stick, before the first pictures were scratched into stone, and long before the first graphic depiction of places on anything like paper, there must have been something we might call premapping: the desire, and so the attempt, to locate oneself."[81] Desire lines can be the path we make or the path we follow, wayfinding and wayfaring, making our way in the world.

LAND LINES

It is good to collect things, but it is better to go on walks.
—Anatole France

"My first work made by walking," explains artist Richard Long, "in 1967, was a straight line in a grass field, which was also my own path going 'nowhere.'"[82] While a student at Saint Martin's School of Art in London, Long wore a single, straight line on a rural field outside of London. His single photograph of the line wore his name into the annals of art like so many footsteps on that field. About the piece, dubbed "A Line Made by Walking," Dieter Roelstraete writes that it "equally belongs to the histories of early Conceptual art, Land art . . . performance or body art . . . and conceptually inflected experimentation in photography."[83] It was Long's first recognized piece of art and set in motion a career that took art out of the gallery and into the landscape. Long says of the piece, "It's made of nothing and disappears to nothing. It has no substance, and yet it's a real artwork."[84]

The line is "a simple metaphor for life," says Long. "A walk marks time with an accumulation of footsteps," he asserts in the 1982 statement "Words After the Fact." "It defines the form of the land."[85] Contextualizing his walking works on a planetary scale, he adds, "The surface of the earth is covered with endless patterns of humans and animals [that have accrued] for millions of years. Each trail and footprint is slightly different, and mine is just one more layer."[86] With these layers, we define our space, our communities, our nations. Ingold writes in *The Life of Lines*, "Walking the roads and paths is to trace a portrait of the country."[87] Roads and paths portray our social and business

Richard Long's "A Line Made by Walking" (1967). © 2024 Richard Long. All Rights Reserved, DACS, London / ARS, N.Y. Reprinted by permission.

connections and their historical ebbs and flows.[88] Human history is made of lines and paths and roads.

As Karen O'Rourke puts it, "contemporary artists have returned time and again to the walking motif, discovering that, no matter how many times it has been done, it is never done."[89] Long claims that walking is his medium of choice.[90] His pieces that aren't created directly by his putting one foot in front of the other are made from things found while walking (mud, stones, etc.), by documenting the conditions of his walks (wind direction, walk duration, length, etc.), or by planning routes on maps beforehand. Sometimes drawing a straight line on a map and then attempting to follow it in the physical world, Long scarcely shares the politics of the Situationists, but his performative use of walking and mapping is quite similar to their ideas of drifting and seeking authenticity in everyday life.[91] Where the Situationists performed the *dérive* in nomadic groups in urban environments, Long walks alone in remote locales. Also, unlike them, Long says that his work is not a reaction to society: "My work is about ideas and actions, it is a balance of the mental and physical."[92] He simply enjoys walking, finding campsites, and just being in the world. The simple spirit of walking, drifting from one point of interest to another, is evident in both.

THE LEY OF THE LAND

Legba . . . was the loa of communication, "the master of roads and pathways."
—from William Gibson's *Count Zero*

Before Richard Long was walking lines in the name of art, some were doing it in the name of the sacred. The "ley" system is a meshwork of lines connecting ancient holy sites. These lines have been observed in several different parts of the world. John Michell describes Alfred Watkins's original discovery of ley lines in Britain in his introductory note to the 1970 reprint of Watkins's *The Old Straight Track* (first published in 1925): "Riding across the hills near Bredwardine in his native country, he pulled up his horse to look out over the landscape below. At that moment he became aware of a network of lines, standing out like glowing wires all over the surface of the country, intersecting at the sites of churches, old stones, and other spots of traditional sanctity."[93]

Watkins's followers found the same phenomenon in South America. Explorer Tony Morrison, who had studied the Nazca lines in Peru, ventured to Bolivia to follow up on a 1932 paper hinting at long and perfectly straight paths. Michell writes, "As first glimpsed by Morrison from the air, . . . the entire landscape was intersected by a pattern of thin, light pathways, running dead straight over all obstacles for distances of more than twenty miles."[94]

Citing advances in aviation and satellite imagery, Michell enthuses, "It has become evident that the ley system in every land is deeply etched into the landscape, being the earliest and most extensive of the layers of markings which the human race has scored on the face of the earth."[95] A patina of age collects and is scraped away, revealing layers of decay. Humans' presence is seen in paths and palimpsests. "The world around us, so much of it our own creation, shifts continually and often bewilders us," writes Kevin Lynch. "We reach out to that world to preserve or to change it and so to make visible our desire."[96]

Where our world and its media used to show the marks of footprints and fingerprints, now their marks are moving out of our hands, in the clouds, in our heads—invisible, ephemeral gestures. Maybe that is the real difference between old and new media: the way they show use. Richard Long once posited, "I think that the surface of the world anywhere is a record of all its human, animal, and geographical history."[97] Whether designed from the top down or emerging from the bottom up, the texture of that history is up to us.

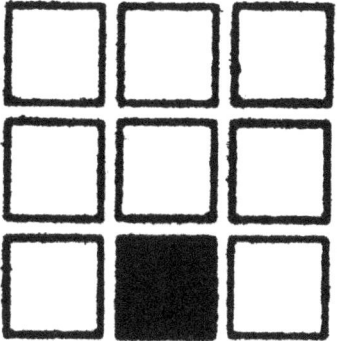

Location Is Everywhere

In the sagas it was said that humans dream with their hands,
only their hands, and so have cities rather than sagas,
monuments rather than memories.
—Peter the dolphin, in Ted Mooney's *Easy Travel to Other Planets*

The city no longer exists, except as a cultural ghost for tourists.
Any highway eatery with its TV set, newspaper, and magazine
is as cosmopolitan as New York or Paris.
—Marshall McLuhan, 1954

There will be no further need for cities or castles.
There will be no further reason for roads or squares.
Every point will be the same as every other.
—Superstudio, 1968

CHRONIC TOWN

Natural places are no different than human cities. The old exists
next to the new. Invasive species integrate with or push out native
species. The landscape you see around you is the same as
seeing an old cathedral next to a skyscraper.

—Ghost Bird, in Jeff VanderMeer's *Acceptance*

Since moving out on my own, I have gravitated toward cities: Seattle, Portland, San Francisco, San Diego, Austin, Atlanta, Chicago. Externalized memories built in brick and concrete. Each time we move to a new city, we make memories as the city slowly takes shape in our minds. Every new place we locate—the closest grocery store, the post office, rendezvous points with friends—is a new point on the map. Wayfinding in a new city is an experience you can never get back. Once you're familiar with the place, it is gone, lost to the entropy of experience. Yet our cities, those densely populated spaces of our built environment, have always been slowly redefining themselves. Italo Calvino's *Invisible Cities* suggests two kinds of cities: "those that through the years and the changes continue to give their form to desires, and those in which desires either erase the city or are erased by it."[1] In 1991 the concept of edge cities emerged, popularized by Joel Garreau.[2] In 2001 we witnessed the worst intentions of a tightly networked community that lacked physical borders—what Richard Norton calls a "feral city," a connected community beyond the reach of traditional law and order.[3]

While design of all kinds is based on the needs of past behavior, it also predicts future use. To interpolate Stewart Brand, all such predictions are wrong.[4] Our capital-driven, networked societies produce, more than anything else, ephemeral things, things built not to last but instead to disappear and be displaced by newer versions. "Cities look to me to be our most characteristic technology," William Gibson says. "We didn't really get interesting as a species until we became able to do cities—that's when it all got really diverse, because you can't do cities without a substrate of other technologies."[5] Some of these include irrigation, agriculture, pottery, metallurgy, literacy, and governance.[6] Cities have been the physical manifestations of our collective concerns.[7] Our thinking about cities needs to evolve as they are evolving. They are perhaps better perceived as what the geographer John R. Borchert calls "settlement machines."[8] We once daydreamed in concrete and steel, but if our collective consciousness is flitting and flickering from one thing to the next, our cities will follow.

Up until the early 1800s, cities were designed for walking, their layout de-

Spike Jonze making marks in the Invisible City.
Photo by Rodger Bridges. © Rodger Bridges. Courtesy of Rodger Bridges.

pendent on how far their citizens were willing to walk to work, shop, socialize, or otherwise conduct their business.[9] Cities sprang up near water. Rivers provided the networks.[10] In the hundred years between 1825 to 1925, the trolley, the subway, the railroad, the steamship, the airplane, and the telegraph enabled the "frictionless space" that shaped our cities, with the metropolis growing outward along their splayed and splintering branches.[11] Cities emerge where supplies and production converge and connectivity is concentrated.[12] Just as the telegraph separated communication from transportation, the current dominant forms of connectivity are not grounded in physical space. John Borchet writes, "The communications revolution has accelerated the weakening of hierarchies, the population redistribution, and the instability that had already seriously undermined the geographical legacy from the railroad era."[13] The cities of the future will emerge with cultural scripts as blueprints, "mattering maps" with landmarks like 9/11, Columbine, Katrina, and COVID-19. Their unity will be condensed from the vapor of memories, all clouds in collective heads, fleeting infrastructure.

In his essay "Garcetti's Bridge to Bicycle Nowhere," Los Angeles writer Joseph Mailander describes the harrowing bike ride across the half-mile Hyperion-Glendale Bridge between "the lands the freeways forgot," Los Feliz and Silver Lake. The traffic signals there currently afford a brief, semi-safe interval between the roaring cars and trucks on the road. "And how are they making this bridge safer?" asks Mailander. "By making the traffic even faster and daring the cyclists to mix with the motorists even more."[14]

Just about everything I have read about urban development has faulted the car for the ills of the city. "A city can be friendly to people, or it can be friendly to cars, but it can't be both," said Bogatá's mayor Enrique Peñalosa while riding a bike through his city in 2007.[15] "The most dynamic economies of the twentieth century produced the most miserable cities. I'm talking about the US, of course—Atlanta, Phoenix, Miami, cities totally dominated by private cars."[16] Bogatá and Peñalosa are the first case study in Charles Montgomery's book *Happy City: Transforming Our Lives Through Urban Design*. Montgomery writes that, as systems, cities are susceptible to self-replication. That is, a design is established, becomes codified in the plans, and spreads itself to other cities. For example, the car-based dispersion that characterizes U.S. cities is encoded in their DNA. "The dispersed city lives not only in the durability of buildings, parking lots, and highways," Montgomery writes, "but also in the habits of the professionals who make our cities."[17]

A disturbing amount of these habits come from military practices. Sophie Yanow's graphic novel *War of Streets and Houses* briefly and beautifully tells a story of struggling with space, place, and the design of both. Of this struggle, she says, "I sat in on an urban planning course once where the professor was talking about how we as a culture in North America have lost a certain 'know-how' when it comes to building and creating spaces. But even if we have the know-how to shape space the way we want to, authority always wants to de-fer to professionals, to urban planners or architects."[18] In *War of Streets and Houses* she cites Foucault's "disciplinary space" to describe the ways urban space is designed to control its inhabitants. Echoing urban theorist Jane Ja-cobs, Yanow continues, "I think that in terms of building social movements, a walkable city is important. Places where people literally brush up against each other on the sidewalk, where they have to be in public together and don't just see each other passing by in cars."[19] Urban space is such a different experience when you're actually in it, on foot or on a bicycle and not in a car or a building. As Rebecca Solnit tells Jarrett Earnest in the *Brooklyn Rail*, "With cities I'm more interested in public spaces and streets, which have been important for my work on democracy and the way that democracy re-quires us to co-exist in public, so I'm more concerned with the space between the buildings than the buildings themselves."[20]

Having grown up in rural Northern California, Yanow first finds downtown Montreal an anonymous space: "Empty. Calm. As if it hid nothing and had nothing to hide."[21] She quickly compares it to places along the coast or in the suburbs where "human scale things are quaint or unimaginable."[22] Democracy happens at human scale. That is why we occupy the streets and not the fields.

In his book *Rebel Cities*, the geographer David Harvey traces the pedigree of urban-based class struggles back to the late eighteenth century. From Paris in 1789 through Paris in 1968, Seattle in 1999, and the more recent Occupy Wall Street in Zuccotti Park in New York City, Harvey situates the city as the center of class struggle. Where others have criticized Occupy Wall Street as unorga-nized and ineffectual, Harvey praises the movement, writing, "It shows us that the collective power of bodies in public space is still the most effective instru-ment of opposition when all other means of access are blocked."[23] But there is less and less public space to fill with bodies. From Georges-Eugène Haussmann in nineteenth-century Paris to Robert Moses in twentieth-century New York, changes in architecture and urban planning may be the most tangible and tenacious result of such political unrest.

Our cities were redesigned to prevent political action and simultaneously reconfigured to accommodate automobiles. Looking ahead, we see more lanes of gridlocked traffic. Mailander writes, "Imagining the future as a cool and pristine place is code for saying things aren't right right now. Some may like to try to fix things by inviting dreamers to dream bigger dreams. But we had better apply some math to these dreams too."[24] Cars drive capital. If we want them out of the city, it is time to learn the algebra of alternatives.

"Nomads rarely, if ever, destroyed a civilisation," writes Bruce Chatwin. "They merely took advantage of a disintegrating situation."[25] And Marshall McLuhan wrote three decades earlier, "Before the huddle of the city, there was the food-gathering phase of man the hunter, even as men have now in the electric age returned physically and socially to the nomadic state. Now, however, it is called information-gathering and data processing."[26] You can see the nomadic spores of future cities evident already in traveling circuses, carnivals, food trucks, pop-up shops, timeshares, home-shares, rideshares, and the nomadic tribes that gather at Black Rock City in northwestern Nevada for the annual Burning Man festival. These are transient topographies, disposable digs, routes instead of roots.

Every fall tens of thousands of people travel to the desert in the northwestern corner of Nevada, accumulating the sixth-largest population center in the state. The welcome sign reads "Welcome to Nowhere." For one week each year, people shed their technological shells and live a simpler existence in the desert. Transportation is primarily provided by walking or riding bicycles, as cars that aren't preregistered for the event are turned away at the gates. Entertainment is provided by ad hoc performances and artworks of various scales and degrees of interactivity. Shortly thereafter the city is fully dismantled, and by the next month evidence of its existence is wiped from the sand.

Like Black Rock, an edge city is one that is perceived by its population as one place. Like in tight-knit neighborhoods, its residents staunchly identify with and defend it, resisting outside influence. Conversely, rapid transit has increased the exchange of ideas between once-isolated places, spurring innovation.[27]

Critiquing the much-lauded coming of the "smart city," the urbanist Adam Greenfield writes that "the important linkages aren't physical but those made between ideas, technical systems, and practices."[28] After all, the first condition for a smart city is a world-class broadband infrastructure.[29] Paul Virilio once expanded William Gibson's term "cyberspace" to an imaginary original form, "cybernetic space-time," the extending of which evokes the ultimate mechanical prosthesis of the mind, a global body, a planet-spanning, command-control system to end all such systems.[30] Somewhere between McLuhan's *global village* and Pierre Teilhard de Chardin's *noosphere*, the seeds are spreading.

The sociologist Saskia Sassen describes global cities as "nodal points for the coordination of processes," as places to produce specialized services, financial innovations, and the making of markets."[31] Since the 1970s there has been a widespread shift from manufacturing to services, from laborers helming machines to workers sitting at computers.[32] This shift mirrors the ones from analog to digital, from muscular to mental. Sassen adds, "The 'things' a global city makes are services and financial goods." These "things" are not physical products. They can be created, managed, and exchanged from anywhere. The top-down hierarchy of the global city can be physically distributed without disrupting any of its functions. Thanks to the internet, the global city can come together from the bottom up.

To use another of Gibson's terms, the online world has "everted" itself into physical space.[33] From flash mobs to terrorist cells, communities can now quickly toggle between virtual and physical organization. How long before our cities do the same?[34] Connection is key—network connection and emotional connection.

Connection could *be* the city.

FROM PURISTS TO TOURISTS

A whale can speak to another whale across sixty miles of ocean.
A whale is as intelligent as we are, just in a way we can't measure or
understand. Because we're these incredibly blunt instruments.
—Ghost Bird, in Jeff VanderMeer's *Acceptance*

We have long been at work, rather unwittingly, building an "ephemicropolis," a sprawling urbanity without roots, reason, or permanence.[35] "We've created 10,000 places that are not worth caring about," James Howard Kunstler, author of *The Geography of Nowhere*, told me in 2002. "Imagine the corrosive effect of that on our national psychology. How soon before we become a land not worth defending?"[36] After World War II, cities were built to accommodate the car, pushing pedestrians aside. Roads, highways, interstates, parking lots, parking garages, drive-throughs, strip malls—these are places for cars, not people. In the decades since, the ephemeral nature of urban development has only gotten worse. In his book *Out of the Mountains*, David Kilcullen defines four global factors that will determine the future of the ephemicropolis: population growth, urbanization, littoralization (the human tendency to cluster along shorelines), and connectedness.[37] As more and more people meet and fall in love and populate the planet, they are doing so in bigger cities, near water, and with more connectivity than ever. Basically the future of human hives is complex, crowded, coastal, and connected.

As soon as the coasts recede with the rising ocean tides, those hives will have to move inland. More importantly, they will have to disassemble, move, and reassemble in some fashion. The urban planner Kevin Lynch once called cities "systems of access that pass through mosaics of territory."[38] That description is fitting for what I call *swarm cities*, a tenuous but slightly more stable form of what the theorist Eric Kluitenberg refers to as "swarm publics": "Today, we are witnessing the rise of swarm publics, highly unstable constellations of temporary alliances that resemble a public sphere in constant flux; globally mediated flash mobs that never meet, fueled by sentiment and affect, escaping fixed capture."[39] Swarms come together when needed and disperse when not. Swarm cities are temporary nomadic tribes made up of like-minded individual monads, a digital diaspora.

Swarm cities are only as physical as they need to be. And, as connected as they are, they are also only as cohesive as their sustainment demands. Cities at a certain scale are shared places of shared experiences and shared concerns. Later and larger, they become sites of separation, divided into different experiences and concerns they once shared.[40] Lynch wrote, "Our senses are local, while our experience is regional."[41]

Meanwhile, Robert J. Sampson argues for behavior based on our idea of local roots at the neighborhood scale. The neighborhood effect is how we describe the interaction between individuals and their main network, and between the

Anish Kapoor's "Cloud Gate," Chicago. Ballpoint sketch by the author.

local and the global. The neighborhood is where boundaries matter. It is where human perception binds us within borders, where nodes are landmarks in a physical network, not connections in the cloud. Like the scratches on a record or hisses on a cassette and the landmarks on the map and our memories of neighborhoods, swarm cities are duplicitous, existing both outside and inside our heads. Early on in his book *In Divisible Cities*, Dominic Pettman repurposes the idea of the "mattering maps," the map we make to and from the things that matter: "A map that generates territory, rather than the other way around. . . . A map that does not represent cities that exist independently, but a map that brings cities into being."[42] In other words, the map that matters is the one we carry in our heads, our cognitive cartography. The participants of swarm cities can share their concerns and experiences across these scales.[43]

Traditional urban planning is top-down, strategic. A tactical approach like those described above is bottom-up.[44] As tactical instead of strategic, ad hoc urbanity also further allows the integration of the natural environment—for example, Google's plans for their new campus in Mountain View, California, in 2015: "The idea is simple. Instead of constructing immoveable concrete buildings, we'll create lightweight block-like structures which can be moved around easily as we invest in new product areas. . . . Large translucent canopies will cover each site, controlling the climate inside yet letting in light and air. With trees, landscaping, cafes, and bike paths weaving through these structures, we aim to blur the distinction between our buildings and nature."[45]

Local communities haven't been diminished by global networks; they have come unmoored because their connectedness isn't physically grounded. "Our electric extensions of ourselves simply by-pass space and time, and create problems of human involvement and organization for which there is no precedent," wrote McLuhan. "We may yet yearn for the simple days of the automobile and the superhighway."[46] Where the Situationists preached the gospel of *dérive* and *psychogeography*, the future of cities needs more of what Laura Oldfield Ford calls *sociogeography*.[47] We need to tune in to the memories lying latent in the landscape, a more connective engagement with the built environment and each other.[48] As Ford's zine *Savage Messiah* illustrated, with each issue focused on a different London postal code, there remains no absolute value to the region you happen to occupy.[49] The nodes shift as needed. This is the long tail of location. This is the edgiest of edge cities. This is the cartography of the future: giant, sprawling mattering maps made of memories. Within them are vast and multiple new swarm cities to explore.

UNWORDED WORLDS

The natural skin of the world was water,
and all water on earth was connected.
—from Omar El Akkad's *American War*

Dolphins are students of the sonic, the tidal, and the
gravitational. Through ear and skin, the dolphin receives forty
million bits of such information per second and organizes them
spontaneously into a changing musical replica of the world.
Some of this music is useful; some is not.
—from Ted Mooney's *Easy Travel to Other Planets*

Marshall McLuhan foresaw telepresence, via electronic media like two-way television and video-telephone, as an antidote to the car-centric city, redefining our sense of place and erasing the separation between work, home, and commerce.[50] Our ideas about media infrastructure begin with the body.[51] Itself a node in multiple networks and composed of networks, the body remains the site of all media. As connected to each other as we might be via external media, we remain embodied, embedded in our environment.[52] Where the bodily boundaries blur is where our media intercedes.[53] Toshiharu Ito writes, "In contrast to human beings who have invented tools to make things and in doing so invented a culture, whales and dolphins may have established a culture without objects, consisting only of communication. Moreover, their cultural structure may have a property which is sympathetic to the new media society which has begun to surround us. This could be indicating the opening of a new civilization that is totally different to the conventional form."[54]

Through the aquatic ambience of the ocean, whale vocalization can be heard over a thousand miles away.[55] Pure connectivity. The history of the internet's spread is largely a story of the breaking down of boundaries.[56] Its environment, such that it is, enables many overlapping connections between what would otherwise be distinct places of business, commerce, recreation, and social activity.[57] The irony of so-called telecommuting is that since physical location no longer matters in the postgeographic workday, it makes it matter even more. The old boundaries are gone.

The irony now is that where we are matters less than the digital wares with which we saturate ourselves. While commuting, at school, at work, at home, on a trip, on vacation, visiting friends—personal media usurps all of these locations with a persistent and precise hold on our attention. "Today, such an information network is gradually permeating into our daily life environment," Ito continues. "In one sense, mankind . . . placed in such an information environment can be compared with the dolphins or whales in a new kind of sea."[58] Lest we forget, Neil Postman's original definition of *media ecology* not only defined media as environments but also the inverse, the environment as media.[59]

"Luxury hampers mobility," writes Bruce Chatwin of the "Nomadic Alternative," the resistance to settling in so-called civilized society and the option of escape to "the steppe."[60] He asserts ten pages later, "All works of art, even mechanical artifacts, reflect the aspirations of their makers, and are eyewitnesses to the past. The art of urban civilization tends to be static, solid and symmetrical. It is disciplined by the representation of the human body and by the mathematical skills attendant upon monumental architecture. To a greater or lesser extent, nomadic art tends to be portable, asymmetric, discordant, restless, incorporeal and intuitive."[61]

If we are to become a pop-up populace, if we are to stay mobile, we can't carry all of our art and artifacts around with us. The culture that will survive will be the one we can carry in our heads.[62]

CHAPTER 9

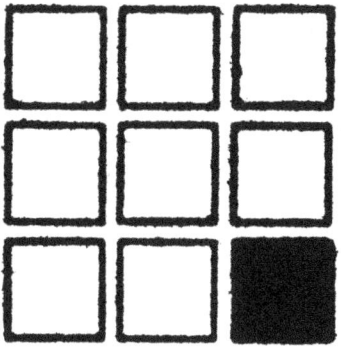

Suture Self

Because we are historical animals, we are always in the
process of becoming, perpetually out ahead of ourselves. Because
our life is a project rather than a series of present moments, we can
never achieve the stable identity of a mosquito or a pitchfork.

—Terry Eagleton, *After Theory*

Nature gave them tongues, technology gave them
loudspeakers, and they all believe that because they can
use both, whatever they say is important.

—Gracie, in Pat Cadigan's "The Final Remake of 'The Return of Little Latin Larry'
with a Completely Remastered Soundtrack and the Original Audience"

Life is a bridge. Cross over it, but build no house on it.

—Indian proverb

DUBIOUS ASSEMBLAGES

Self-mediation is the act of constituting presence in a mediated environment.
Formerly a marginal practice it has now moved to center stage.

—Eric Kluitenberg, *Delusive Spaces*

The first time I saw Laurie Anderson was on *Saturday Night Live*. I will never forget that night, standing in my parents' living room aghast. Her spiky hair, pitch-shifted voice, the wild stories she told. She performed the songs "Beautiful Red Dress" and "The Day the Devil." One was about gender inequality and the other about the devil coming to get you. At fifteen, I was as intrigued as I was terrified. When her next record, *Strange Angels*, came out, I got it on cassette, CD, and LP. I don't know why. I just had to have it on all available formats. According to McLuhan, our media exist at the intersection of technologies and bodies.[1] Anderson has been performing on that corner since the 1970s.

Trained as a violinist from age five, she adapted and eventually abandoned the instrument. In 1975 she constructed a tape-bow violin. A cassette playback head was mounted on its bridge, and horsehair on the bow was replaced by a strip of magnetic audiotape. The sounds on the tape play forward or backward depending on how the bow is dragged across the head.[2] Among the recordings on the tape-bow were human voices, animal cries, and bits of music—everyday sounds Anderson manipulated into bizarre chirps and scrapes and stretched into yawning statements. "I wanted to make songs that were more like remembering than listening," she says.[3] Anderson claims she was using the violin as less of an instrument and more of a prop, "more as a ventriloquist's dummy."[4] In the process, she was also inventing a new kind of instrument that merged tradition with transgression in technological form. Almost a decade later, Christian Marclay's early performances would feature his phonoguitar, a guitar with a turntable in place of strings.[5]

Starting her art practice as a sculptor, Anderson quickly moved on to performance, music, film, and more recently virtual reality. During that transition, she landed an unlikely pop hit. The eight-minute, twenty-one-second "O Superman" rose to number two on the UK singles chart in 1981.[6] The song, which loops several simple elements, primarily Anderson's voice, a keyboard, and a sample of chirping birds, creates a lush, repetitive, and minimalist soundscape, and it sounds as fresh and futuristic now as it did in the early 1980s. It is both a critique of the U.S. foreign policy at the time and a cross between

Laurie Anderson in *Home of the Brave*, 1986. Licensed from Alamy Stock Photo.

the banality of a message from your mother on an answering machine and the distant early warning of an aerial attack. Mark Dery describes Anderson's work as meditating on mediated society, using phrases like "virtual vaudeville" and words like "cybercabaret."[7] He describes Anderson's six-hour performance piece from 1984, *United States, Parts I–IV*, as "Marshall McLuhan's *The Medium Is the Massage* set to music."[8] That book, a collaboration between McLuhan, Quintin Fiore, and Jerome Agel, furthers many of McLuhan's probes through a typographical explosion of text and images that stretches both the medium and the mind. Anderson's *United States* is about mental states and emotional states as much as it is a nation of states, after all.[9]

"As an artist, I have always tried to connect two worlds," she says, "the so-called real world and the other world, an alternate world of possibility and chance: a dream world."[10] Her performances blur those lines further. In her 1986 concert film *Home of the Brave*, her body, augmented by technology on stage, by turns clashes and meshes with her very presence, dissolving into the video background.[11] Other times her shadow looms over her body from behind, giving her the appearance of an outsized specter. The film brings together her invented instruments—most famously the tape-bow violin and the drum-set bodysuit—computer animation, background video, and vocal experiments, the latter of which find Anderson fading into the machinations: technological possibilities always tempered by the threat of war and self-destruction.

"When you have too much equipment, that's a real limitation," she says, echoing our concerns from earlier. It's intriguing that one of the early adopters of advanced recording equipment find them limiting in their abundance. Acknowledging the irony, Anderson says, "It's a strange sort of paradox. I use old electronics as well. I like trying to push them into other modes—use them for things they were not supposed to be used for. That's satisfying."[12] The affordance mining of zine-making, skateboarding, and desire lines are all evident in Anderson's attitude. "But you can get into a trap," she continues, "I think a lot of musicians are in that trap—trying to get the latest thing that's going, to fix everything up. It won't."[13] Seduction is compromise, and technology tempts us to use it beyond our needs. It is great if you learn its limits and don't lean on it like a creative crutch. The world is limitless. Our time and attention are not. Our tools shouldn't be either.

"My biggest fear is being isolated," Anderson says. "My biggest fear is losing contact with other people, forgetting how to express things or just suddenly not being able to communicate at all."[14] Fortunately, Anderson is a really good communicator. She once performed an impromptu lecture at a New York college, making up stories to go along with the slides provided.[15] "I think that's a big function of stories: survival," she says. "Tell a story that lets you live—whether it's your own story or somebody else's story, it allows you to live."[16] Telling stories requires communication, sharing a piece of yourself—of your attention—with someone else.

"It all became very wide open from the moment John Cage said that everything sounded as good as everything else," Anderson says, "and it was only our limitations that made it less beautiful."[17] Like Cage, she bristles at the stasis of recorded work, preferring the ever-evolving environment of performance.[18] "I think the thrill of creation is making something from nothing. Yet whenever I stop to listen, for just one moment, to whatever is happening in this giant ocean of sound, I thank John Cage for making me pay attention. There are some things you don't really need to manipulate. Art is about paying attention."[19]

Moreover, Herman Rapaport writes: "Anderson's work is not part of a historical forgetting, but is an attempt to describe accurately through pastiche how the articulation of vernacular and elite cultures manages to suture the historical subject. This suturing paradoxically elicits experiences of a decentered and detotalized consciousness whose nostalgic, vernacular expressions reveal an uncomfortable alliance between elite and vernacular cultures."[20]

The easy consumption of radio and film forced the distinction between high and low culture based on the quality of attention each required. As Walter Benjamin put it, "The masses are criticized for seeking distraction in the work of art, whereas the art lover supposedly approaches it with concentration."[21] "Rather than searching the wreckage of culture for depleted bits of language," Rapaport writes, "Anderson looks for expressions which resonate in terms of the postmodern relations between vernacular and elite culture." Whereas other performers are what Gene Youngblood called "merchants of mannerisms," imitators repackaging pop cultural "data and phenomenon," Anderson merges high and low cultures, running counter to so-called "nobrow culture."[22] Like the personal-as-political post-punk of Gang of Four, she uses her art to critique everyday life.

Laurie Anderson has done as much as anyone to explore the place of the self in the face of extreme technological mediation.[23] "Everything I've made has been about disembodiment," she says.[24] "I am in my body the way most people drive in their cars," she adds elsewhere.[25] Her work constantly simulates possible presents in which identity based on the body and gender are undone and reassembled by technological projections. Timothy Morton writes, "Anderson's voice provides a taste of something that is disturbingly just 'there.'"[26] The closer we get to her, the further she slips into disembodied voices, sounds, and images. She enacts the simulations she critiques. "Anderson's body is not one," writes Rosi Braidotti, "but a shifting horizon of technologically mediated transitions: an acoustically gifted cyborg."[27] Anderson is "fragmented, dispersed, and decentered."[28] It is as if she is not really there at all. It is as if she is an assemblage of appendages, a person made of media.

"Because the self is partly a product of its communications," Erik Davis writes, "new media technologies remold the boundaries of being."[29] Those boundaries as demonstrated by Laurie Anderson's performances represent a system of assemblages and flows, much like those found in the insect world. Jussi Parikka notes that when we compare media as the extensions of humans, as McLuhan and others have, to media as the externalized world of insects, we run into severe problems when it comes to certain technologies, namely wheels and fire.[30] He writes that we must stop thinking about bodies as closed systems and realize that they are open and constituted by their environment.[31] Our skin is not a boundary; it is a periphery: permeable, vulnerable, and fallibly open to external flows and forces. "We do not so much have media as we are media and of media," Parikka writes. "Media are brains that contract forces of the cosmos, cast a plane over the chaos."[32] We can no longer do without, if we ever could.

ENNUI GO

The One has no meaning. It is an abstract, unintelligible entity.
One thing is nothing. One single person is no one.
—Jean Baudrillard, *Cool Memories, IV: 1995–2000*

At present we think of science and technology as means of mastering
the world. But the self that struggles to master the world is only a
shimmer on the surface of things.
—John Gray, *Straw Dogs*

"Irony used to feel like a defense against getting played," writes the novelist Hari Kunzru, "a way for a writer to ward off received ideas and lazy thinking." Broadly speaking, irony is the rhetorical strategy of saying one thing yet meaning another, usually the opposite. It also might be the most abused trope of our time. It's gone beyond a ratio of *substance* over *style* to one of *absurdity* over *authenticity*. "It also made us feel nihilistic and defeated," Kunzru continues. "More recently we've seen how it can be a screen for reactionary politics."[33] In the preface to his 1999 book *For Common Things*, Jedidiah Purdy frames the overbearing irony of our era as a defense mechanism: "It is a fear of betrayal, disappointment, and humiliation, and a suspicion that believing, hoping, or caring too much will open us up to these." Irony is an escape route, an exit strategy, a way off the hook in any situation. It has become the dominant mode of pop culture, and we're all tired of it.[34]

In his book *The Comedian as Confidence Man*, Will Kaufman explains the feeling, what he calls *irony fatigue*, the exhaustion of ironic distance as the promise of play collides with the pursuit of truth. He discusses the comedian Bill Hicks having to edit lines from his unaired twelfth appearance on *Late Night with David Letterman*. Hicks maintained his "Warrior for Truth" persona, claiming all the while that his words were "just jokes." He didn't mean to offend; he was just kidding.[35] Having it both ways is perhaps impossible for a figure under public and media scrutiny, but what about your classmates? What about the coffee shop denizen? Are they for real, or are they joking? Why is everyone so veiled in irony? Princeton professor Christy Wampole writes, "Ironic living is a first-world problem. For the relatively well educated and financially secure, irony functions as a kind of credit card you never have to pay back. In other words, the hipster can frivolously invest in sham social capital without ever paying back one sincere dime. He doesn't own anything he possesses."[36]

Three major cultural epochs came and went in the meantime: cool became uncool, the nerds had their revenge, and stark sincerity was pushed to its breaking point. One was already faltering when the twenty-first century arrived. Everything that used to be cool is now remade, rebooted, or recycled. Resorting to irony is the only response that quells the cognitive dissonance of dealing with such a contradictory world.[37] Between the death of cool and the ironic now, the geeks rose to rule all, and emo culture came to the fore, the latter allowing young men to reveal their emotions. We all know the story of the geeks.[38] Theirs was a rise to riches, an underdog having its day, but the emo kids never enjoyed such empowerment.

In America's post-9/11 cultural climate of mourning, confusion, anger, and uncertainty, the emo subculture gained momentum as a way for young people to express and deal with their anger and uncertainty. The music and the open wounds allowed young people mourn in public. In his book *Nothing Feels Good*, Andy Greenwald frames emo culture as a teen phenomenon, a culture of kids who haven't "thought the deep thoughts yet—they're too caught up in their own private drama and they've found a music that privileges that very same drama—that forces no difficult questions, just bemoans the lack of answers."[39] Post-9/11 America might have been about forcing the difficult questions, but it was just as much about bemoaning the lack of answers, and emo made either one okay. Coming of age already leaves teenagers feeling uprooted and untethered, with no home and no sense of belonging. The feeling was exacerbated by the events of September 11. Not only were their bodies and relationships changing in unprecedented ways, the world seemed to be doing the same thing. As Robert Pogue Harrison puts it, "Wherever the real imposes itself, it tends to dissipate the fogs of irony."[40] This lack of roots provides the backdrop for the mass emergence of emo culture. Emo allowed dudes to be as sappy and sincere as they wanted to be. "If we stay with the sense of loss," Judith Butler writes, "are we left feeling only passive and powerless, as some might fear?"[41] The feeling of being only passive and powerless is at the core of emo culture. Butler continues, "Or are we returned to a sense of human vulnerability, to our collective responsibility for the physical lives of one another? Could the experience of a dislocation of First World safety not condition the insight into the radically inequitable ways that corporeal vulnerability is distributed globally? To foreclose that vulnerability, to banish it, to make ourselves secure at the expense of every other human consideration is to eradicate one of the most important resources from which we must take our bearings and find our way."[42] Where emo culture folds under the weight of affect and uncertainty, Butler urges us to follow it outward. *Parks and Recreation* and *The Good Place* creator Mike Schur says, "Sincerity is the opposite of 'cool' or 'hip' or 'ironic.'"[43] But if our pop culture is just recycling plastic pieces of the past, where it is dragging us?[44]

Simon Reynolds draws a parallel between nostalgic record collecting and finance, "a hipster stock market based around trading in pasts, not futures," in which a crash is inevitable.[45] "The world economy was brought down by derivatives and bad debt; music has been depleted of meaning through derivatives and indebtedness."[46] After all, what is emo if not punk rock peanut butter dunked in goth chocolate? For better or more likely for worse, what emerged after emo culture was the cult of irony. In the ennui of the everyday, we no longer strive to be sincere or cool but instead coldly ironic. The "corrosion of nostalgia" as William Gibson calls it, might leave us longing for simpler times, but they are times not taken to heart.[47] Filters on digital photos that make them look old represent not only longing but also the undermining of that longing. It is irony fatigue filtered in sepia and framed like a Polaroid.

To live in the image of irony is to avoid risk.[48] It means not ever having to mean. If you meet someone who truly lives unironically in the moment, the only thing that matters is that moment. Value only accrues over time. That is, things matter in their historical contexts. Sincerity is only a salve to irony if it is given unmediated room to breathe, but irony holds us hostage, mired in the current moment, and we carry our personal media that carry their own ironies.

PRESENT TENSION

What you don't like are sealed black boxes. Okay, let's open them. Once we see inside, not only do things get a lot less scary, we'll learn. Learn amazing new things.
—Henry Capaldi, in Kazuo Ishiguro's *Klara and the Sun*

When I was growing up, the year 2000 was the temporal touchstone everyone used to mark the advances of modern life. Oh, by then we would be doing so many technologically enabled things. Cars would fly and run on garbage, computers would run everything, and school wouldn't exist. We were all looking forward, and Y2K gave us a point on the horizon to measure it all by. When it came and went without incident, we were left with what we had in the present. Rosi Braidotti writes, "The fast turnover of available commodities and the acceleration they induce results, in fact, in our being in a state of constant jet lag; we are structurally always behind the times and getting synchronized is a real challenge. . . . The same logic of titillation without ever reaching fulfilment is at the heart of contemporary culture . . . such as the *Star Wars* series. These are legalized but forceful forms of mild addiction. Being kept hanging is not only addictive, it is also intrinsically frustrating."[49]

Chasing the latest new device, running the treadmill of software upgrades, swimming the infinity pools of social media: so it goes for many in the twenty-first century. "If only we could catch up with the wave of information, we feel, we would at last be in the now," Douglas Rushkoff writes.[50] "We spent centuries thinking of hours and seconds as portions of the day," he says, "but a digital second is less a part of greater minute, and more an absolute duration, hanging there like the number flap on an old digital clock."[51] A digital clock is good at accurately displaying the time right now, but an analog clock is better at showing you how long it's been since you last looked. Needing, wanting, or having only the former is what Rushkoff calls "a diminishment of everything that isn't happening right now—and the onslaught of everything that supposedly is."[52] As the song goes, "When you say it's going to happen now, when exactly do you mean?"[53]

We are as sieves, filtering news from noise, or as sponges, soaking up information and influence of all kinds. We are not so much immersed in media as we are saturated by it. The psychologist Kenneth Gergen writes, "Emerging technologies saturate us with the voices of humankind—both harmonious and alien. As we absorb their varied rhymes and reasons, they become a part of us and we of them. Social saturation furnishes us with a multiplicity of incoherent and unrelated languages of the self."[54] Ours is a chronic presence in a chronic present.[55] Our online profiles give us an atemporal agency whereon we are there but not actually present. On the other side, our technologies mediate our identities by anticipating or projecting a user.[56] Walls, fences, locked doors, online moderators—"the doormen of discussions," spam filters, and other gate-keeping contrivances protect the private from the public and vice versa.[57] Even with such boundaries in place, our embodiment is still at risk. Our lives are so saturated, so mediated by technology, that we have to turn it off to be present in the present.[58] Truncated and clipped, we shrink to fit the roles our media allow, and some of us find them more comfortable than others.[59]

In the corporate-control culture, set off by information technology and the mainframe computer, business managers came to be seduced by the system. They used the abilities of their information systems to keep tabs on their workers—even where there had previously been no apparent problems. In her 1997 tech memoir, *Close to the Machine*, Ellen Ullman describes installing a computer system in an insurance agent's office: "The company's employees had been there for ten and twenty years, particularly the women, mostly clerical workers. They were the ones who would be most affected by the new system, yet they went about learning it with a homey cheerfulness that surprised me."[60]

Their manager asked Ullman if the system could record their keystrokes. He wanted to leverage the data-accumulating abilities of the new system to find out what his employees were doing, not because there was a problem with their productivity but instead because it was now possible to do so. He had "succumbed to the fever of the system," as Ullman puts it.[61] We get seduced by the possibilities of the system even if those possibilities are beyond our needs. "Many years and clients later," Ullman continues, "this greed for more data, and more again, had become a commonplace. It had become institutionalized as a good feature of computer systems: you can link them up, you can cross-check, you can find out all sorts of things you didn't set out to know."[62] It's no wonder then that our personal media knows much more about us than we might like or have originally anticipated. While the possibilities and power of the server side of the system are seductive, those same forces shape users' behaviors on the client side. And we act accordingly.[63]

YOU ARE HERE

The machine is not the environment for the person; the person is
the environment for the machine.
—Aviv Bergman

The long-range question is not so much what sort of environment
we want, but what sort of people we want.
—Robert Sommer, *Personal Space*

"How did you get here?" asks Peter Morville on the first page of his book *Ambient Findability*.[64] It's not a metaphysical question but rather a practical and direct one. Ambience indirectly calls attention to the here we are in.[65] It is always all around us. Timothy Morton's *The Ecological Thought* explains it this way: "Take the music of David Byrne and Laurie Anderson. Early postmodern theory likes to think of them as nihilists or relativists, bricoleurs in the bush of ghosts. Laurie Anderson's "O Superman" features a repeated sample of her voice and a sinister series of recorded messages. This voice typifies postmodern art materials: forms of incomprehensible, unspeakable existence. Some might call it inert, sheer existence—art as ooze. It's a medium in which meaning and unmeaning coexist. This oozy medium has something physical about it, which I call ambience."[66]

"Ambience" is a loaded little word at best. In his book *Ambient Commons: Attention in the Age of Embodied Information*, Malcolm McCullough reclaims the word for our hypermediated surroundings, saying that we have mediated aspects of our world so well that we have obscured parts of the world. Looking through the ambient invites us to think about our environment—built, mediated, situated, or otherwise—in a new way. McCullough asks, "Do increasingly situated information technologies illuminate the world, or do they just eclipse it?"[67] As Laurie Anderson quips, "Sometimes one medium is too many."[68] How much can we augment before we begin to obstruct? How far flat can we press the extremes of our world?

Canadian theorist Arthur Kroker once described the mediated spirit of the

1990s as a *spasm*: everything floating and flickering, oscillating and isolating. Bruce Sterling writes, "What . . . Kroker calls 'spasm' is a state when you feel totally hyper and nauseatingly bored. That gnawing sense that we're on the road to nowhere at a million miles an hour."[69] Since the 1990s, the feeling has expanded via media technology: invisible networks and ubiquitous screens, social media, mobile devices. In response to this environment, a weird, detached irony has become our default emotional setting: an all-consuming cultural state that flattens everything without regard to the verve or variety of its constituent parts, simultaneously exciting and numbing our senses.[70] It is information anxiety coupled with abject boredom.[71] What happened to the chasm between those two extremes?

In her song, "The Language of the Future," Laurie Anderson says,

> Always two things
> switching.
> Current runs through bodies
> and then it doesn't.
> It was a language of sounds,
> of noise,
> of switching,
> of signals.
> > It was the language of the rabbit,
> > > the caribou,
> > > the penguin,
> > > the beaver.
>
> A language of the past.
> Current runs through bodies
> and then it doesn't.
> On again.
> Off again.
> Always two things
> switching.
> One thing instantly replaces
> another.
>
> It was the language
> of the Future.[72]

Anderson calls this toggling of opposites a "system of pairing." That binary belies a bulging, unexplored midsection. Similarly, William Gibson writes, "This perpetual toggling between nothing being new, under the sun, and everything having very recently changed, absolutely, is perhaps the central driving tension of my work."[73] The space between that switch from one extreme to the other obscures an important aspect of technological mediation and our current state of being.[74] "It's John Cage's birthday," writes Anderson on September 5, 2003. "We listen to Cage reading from his own work. He tells the famous story about when he was in an anechoic chamber, a completely silent room. And he heard two sounds. One was high and one low. As it turned out, the high sound was his nervous system, and the low sound was his blood."[75] Two more extremes we live in between. Two more extremes with ambience in between.

In 1994 Hakim Bey wrote, "All experience is mediated—by the mechanisms of sense perception, mentation, language, etc.—& certainly all art consists of some further mediation of experience."[76] Mediation isn't something we can eradicate any more than we can exist without technology. The word *ambient* describes a situation when two or more things are harmonized into one.[77] We find our mediated selves by pulling these extremes apart. As with a cramped muscle, the solution to Kroker's metaphorical *spasm* is to stretch it out. All the way out.[78] "Try to do something on the right scale—something that you can do yourself," Laurie Anderson offers as an antidote. "Very dangerous art can be made with a pencil."[79] The DIY methods of punk culture are always available, the technology always gets cheaper, and the artifacts of culture are always there to be manipulated.[80]

In his 2010 book *Soul Mining*, Daniel Lanois describes his first band sharing an amp. A single amplifier with multiple inputs was all they could afford. He writes, "The speaker did its best to accommodate the transients, the sudden loud sounds that momentarily poked out to feature a dynamic moment, as expressed by the musicians. Not all expressions at once, only the loudest expression at a given moment."[81] We watched a similar phenomenon play out in real time through the choppy live feeds of protesters of Occupy Wall Street in Zuccotti Park, as they responded to guest speakers in waves so that everyone there could hear the message. This call-and-response is called "the human microphone" and was used due to restrictions on amplified sound in the public space of New York City. It has been used in previous gatherings in lieu of permits for bullhorns and other amplification. In an ironic mix of collaborative leadership, collective allegiance, communication technology (and lack thereof), the human microphone, like Lanois's one-point source, is the perfect metaphor for the distributed, mediated self. "This technological limitation offered something beautiful," Lanois continues. "A one-point source is a dynamic delivery of a group effort. In its primal way, it did away with competition."[82] Now it seems competition is all we have.

"Ambience points to the here and now," Tim Morton writes. "In a compelling way that goes beyond explicit content. . . ambience opens up our ideas of space and place into radical questioning."[83] Just as metaphor calls attention to language, ambience calls attention to place, the here we are in and the now we are experiencing, bringing the background into the fore. "It was like constructing geography and not populating it," Brian Eno writes of his ambient soundtracks. "The listener, I felt, became the population of a sonic landscape and was free to wander round it."[84] Ambience opens up our ideas about everything into radical questioning, out here at the very edge of culture, at the tip of the longest tail, where everything sprays into a mist, threatening to evaporate.

We must make sure that vapor condenses around us in a manner that serves us and not the media.[85] When we and the media merge, we need to make sure that what is left is what we want. Take the extremes of anxiety and boredom—the fog of nothing changed and everything new—and stretch them into an inhabitable environment, into a now worth knowing.

ACKNOWLEDGMENTS

I recently got an email celebrating the twenty-first anniversary of my long-abandoned LiveJournal account. I looked back at my six entries from 2002 and found the seeds of *The Medium Picture*: early books I read on media theory, a note on Brian Eno's *edge culture*, the claustrophobia I felt from working on computer screens.

This was supposed to be my first book. I started outlining it in 2001, worked with an agent for a while, and—after a decade of research and revision—I signed a contract for it in 2011. The book then went through several publishing shuffles, during which I went on to finish several other projects. I worked on it off and on in the meantime and am happy to finally have it out of my head and into your hands.

Many thanks to the early readers of this material. Douglas Rushkoff, Howard Rheingold, Erik Davis, McKenzie Wark, Steven Shaviro, Alex Burns, Matt Schulte, Mark Wieman, David Patterson, David Barker, David Miller, Kasey Pfaff, Bill Minutaglio, Matt Bialer, Peter Relic, Priya Nelson, and Doug Armato all read early drafts and gave invaluable comments and criticisms. Mark Dery, Leo Hollis, Michael Schandorf, and Nicole NeSmith deserve special mention for their extra thorough reads and feedback. Thanks to Paddy McCaffrey-Allen, Tim Ingold, Colin Renfrew, Richard Long, Christian Marclay, Jeff Nicholson, Liliana Porter, Rodger Bridges, Grant Brittain, Eben Sterling, Pat Graham, and Tim Saccenti for helping me with photos and images. Thanks to Ian MacKaye, Jenny Toomey, Miki Vuckovich, Dave Allen, Tod Swank, Eben Sterling, Mike Burnett, Jaime Meline, Duane Pitre, and Calvin Johnson for taking the time to talk with me. Extra special thanks to Andrew McLuhan for his thoughtful and generous foreword.

Every book is a network. Among the helpful nodes in this one were Claudia Dawson, Jussi Parikka, Paul Levinson, Dominic Pettman, Kristen Gallerneaux, Dave Tompkins, Timothy Baker, Ken Jordan, Ashley Crawford, Katie Arens, Joshua Gunn, Barry Brummett, Matt McGlone, Jeffrey Sconce, John Oakes, Brian Johnson, David Silver, Brian Spitzberg, Martha Lauzen, Zizi Papacharissi,

ACKNOWLEDGMENTS

Steve Jones, Richard Nash, Matt Browne, John David Smith, Etienne Wenger, Matt Gold, Steven Johnson, Laurie Anderson, Brian Eno, Bruce Sterling, William Gibson, Annalee Newitz, Cory Doctorow, R. U. Sirius, Matt Gentling, Brian Tunney, Justin April, Chad Foreman, Erik Ellington, Hassan Abdul Wahid, Andy Jenkins, Mark Lewman, Spike Jonze, Dave Sardy, Etienne Turpin, Charles Yu, Tim Maughan, Will Wiles, Gary J. Shipley, Drew S. Burk, Robin Mackay, Charlotte Gusay, Saul Williams, William Hutson, Mars Kumari, Suraj Patra, Dan Barrett, Lance Caldwell, Kaya Oakes, Kyle Beachy, Roisin Kiberd, Peter Bebergal, Kodwo Eshun, Charles Mudede, Craig Gates, Patrick Barber, and Anondi King.

Extra special thanks to Nate Holly, Beth Snead, and Mick Gusinde-Duffy at the University of Georgia Press for believing in this book and ushering it into the world.

Lily Brewer has been listening to me talk about the ideas in this book since we met over a decade ago, so she deserves special thanks not just for tolerating that and everything else I subject her to, but also for helping clarify much of my thinking about them.

PUBLICATION CREDITS

Part of chapter 4 was published as "The Meme Is Dead, Long Live the Meme" in Alfie Bown and Dan Bristow's book, *Post Memes: Seizing the Memes of Production* (Punctum Books, 2019). Parts of the *mining affordances* section of chapter 5 first appeared in the Summer–Fall 2024 issue of the *Henry Ford Magazine*. Parts of chapter 8 about swarm cities first appeared on Steven Johnson's website *How We Get to Next*. And a section of chapter 9 was published by *Sublation Magazine* in 2023.

CHAPTER HEADING IMAGE CREDIT

The 3x3 square design used in the chapter headings was designed by Ameet Hindocha for the 3by3 Music label and the band Cloaks. (You should go find their stuff immediately.) Many thanks to Ameet and Steve Harris for letting me repurpose it here.

NOTES

PREFACE

Epigraphs are drawn from the following sources: Ballard quoted in Mark Dery, "J. G. Ballard's Wild Ride," in *Extreme Metaphors: Collected Interviews*, ed. Simon Sellars and Dan O'Hara (London: Fourth Estate, 2012), 342. Broeke quoted in Harold A. Innis, *The Bias of Communication* (Toronto: University of Toronto Press, 1951), xxvii. Burroughs quote from *Naked Lunch* (New York: Grove Press, 1959), 22. Gray quote from *Straw Dogs* (New York: Farrar, Straus, And Giroux), 2002, 14. Kahn quoted in Gin S. Malhi, "Quality of life . . . on Earth," *Acta Neuropsychiatrica* 22 (2010), 1.

1. Throughout this book I use the terms *media* and *medium* in reference to "a middle, intermediary state" (Gleick, *The Information*, 2011, 153) and, more thoroughly, as "socially realized structures of communication, where structures include both technological forms and their associated protocols, and where communication is a cultural practice, a ritualized collocation of different people on the same mental map, sharing or engaged with popular ontologies of representation" (Gitelman, *Always Already New*, 2006, 7).

2. Solnit, *River of Shadows*, 2003; Muybridge, *Animals in Motion*, 1957.

3. According to Anthony Wilden, "'Mediate,' which used to be a common technical term, means 'not immediate'. 'Mediation,' in its logical and scientific sense of a medium, channel, agency, means, third term, or other indirect relationship by which two or more subjects, objects, patterns, systems, or other relationships are connected or communicate with one another, first went out of style with the nineteenth-century legacy of Hegel, the 'philosopher of mediation'. . . . Since the 1960s mediation has become a technical term more and more favored by neurologists in discussing the communication and control systems of the body, notably the systems of intercellular communication within and between the central nervous system and the hormone system." Wilden, *The Rules Are No Game*, 1987, 160–61. See also Rasmussen, *Communication Technologies and the Mediation of Social Life*, 1996, 19–22, 35–37.

4. Peele, the narrator of the old horror show's reboot, has his own connections to Sterling's *Twilight Zone*, and, like the rest of his movies, *Nope* has its own nods to the nerds (e.g., a *Northern Exposure* hat, O. J.'s "Circle J" horse trailer, 6:13, Angel's Earth, and the Jesus Lizard and Rage Against the Machine T-shirts).

5. Patterson, *The Vast of Night*, 2019.

6. Peele, *Nope*, 2022.

7. Quoted in Kunzru, "Attention," 2021.

8. Thanks to Dr. Martha Lauzen for sharing this anecdote about John Naisbitt's research.

9. This methodology has been outmoded in the meantime, as the internet hammered flat the differences in news coverage across the country. See Klein, "I Didn't Want It to Be True," 2022.

10. I include this caveat because of my frustration when reading David Wortley's *Gadgets to God: Reflections on Our Changing Relationship with Technology* (2012), in which he seems to assume everyone has had the same experience he did with every technological advance discussed.

11. Gray, *Straw Dogs*, 2002, ix.

12. If you are interested in how the cover image came about, I put together a little photo-essay of the process: "The Medium Picture Object Thing: A Photo History," October 3, 2021, https://www.themediumpicture.com/the-medium-picture-cover-story.

13. Coupland, *Marshall McLuhan*, 2010, 165.

14. See Wolf, "The Wisdom of Saint Marshall," 1996, 122–31, 182–87; Marchland, *Marshall McLuhan*, 1989, 43; Dery, "The Mechanical Bridegroom Stripped Bare," in Strate and Wachtel, *The Legacy of McLuhan*, 2005, 95–106.

15. Outside of updating some of McLuhan's extensive work, I also found that my three stages of technological mediation—*separation*, *threshold*, and *crossing*—closely resemble Arnold van Gennep's three stages of ritual rites of passage. I found this more comforting than anything else. The two sets of processes are very similar, and, like McLuhan's, van Gennep's ideas have become the very vocabulary we use to talk about his area of study.

16. Greil Marcus, *Lipstick Traces*, 1989, 414.

17. Culkin, "A Schoolman's Guide to Marshall McLuhan," 1967, 51–53, 70–72.

CHAPTER 1. ERR APPARENT

Part 1 epigraphs are drawn from the following sources: Frisch quotes from *Homo Faber: A Report* (London: Abelard-Schuman, 1959), 178. Harakaway quote from *Angelmaker: A Novel* (New York: Knopf, 2012), 59. Temkin quoted in Rory Litwin, *Speaking of Information* (Duluth, Minn.: Library Juice Press, 2009), 21.

Chapter 1 epigraphs are drawn from the following sources: Baudrillard quote from *Fragments: Cool Memories III, 1991–95* (London: Verso, 1997), 34. Eco quote from *Travels in Hyperreality: Essays* (New York: Harcourt Brace Jovanovich, 1986), 146. Lovink quote from *Dark Fiber* (Cambridge, Mass.: MIT Press, 2003), 10. Dick quote from "How to Build a Universe That Doesn't Fall Apart Two Days Later," in *I Hope I Shall Arrive Soon* (New York: St. Martins, 1985), 9. Pynchon quote from *Vineland* (New York: Little, Brown, 1990), 38.

1. Genette, *Paratexts*, 1997, 1–2. Max Dawson adds, "As is the case with literary works, screen media are surrounded by dense accumulations of paratexts, which in the case of television programmes may include promos, opening credit sequences, websites and vast quantities of user-generated media." Dawson, "Television Abridged," 2011, 41. Brian Eno writes, "Acknowledging that a computer is actually a place for sticking Post-it notes (mine is surrounded by them), make the frame much bigger—give the conceit 'desktop' some real meaning. The problem with computers is that they exist too exclusively in the electronic realm: what you need is a transitional area around the edge." Eno, *A Year with Swollen Appendices*, 1996, 9.

2. Quoted in Scott, "Asking Cosmic Questions," in Kelly, *The Donnie Darko Book*, 2003, xvii.

3. See "Weber's Law of Just Noticeable Differences," USD Internet Sensation & Perception Laboratory, University of South Dakota, https://web.archive.org/web/20141015041433/http://apps.usd.edu/coglab/WebersLaw.html.

4. For a full exploration of this idea, see Evens, *Sound Ideas*, 2005. McKenzie Wark writes, "Air has so much to say for itself. Sound is just bugged air." Wark, *Dispositions*, 2002, n.p. Siegfried Zielinski adds, "An interface separates things, or the concept would make no sense. An interface connects things, or the concept would make no sense either. An interface marks a difference." Zielinski, *Variations on Media Thinking*, 2019, 49.

5. James Carey wrote of the telegraph, "It permitted for the first time the effective separation of communication from transportation. This fact was immediately recognized, but its significance has been rarely investigated." Carey, *Communication as Culture*, 1992, 203. See also McLuhan, *Understanding Media*, 1964, 97.

6. Spigel, *Make Room for TV*, 1992, 1. For comparison, the previous major media technology, radio, took two decades to spread to 80 percent of American homes. See Bagdikian, *The Media Monopoly*, 1983, 197. In Mark Neale's 2000 documentary *William Gibson: No Maps for These Territories*, William Gibson remembers television's arrival like this:

> The only memory I have of a world prior to media is of standing in a peanut field on a farm in Tennessee, looking down the hill at a black, 1950s, sort of, late '40s panel truck, driving along the road.
>
> One of the next earliest memories is of my father bringing home this wooden, box-like thing, with a cloth grille on the front, and a little round, circular television screen, which, I believe, we had for some time prior to there actually being any broadcast to receive.
>
> And then there was a test pattern. I think the test pattern preceded any actual broadcast for several weeks, and the test pattern itself was only available briefly, at scheduled times. And people . . . neighbors, would come, and they would look at this static, non-moving pattern on the screen that . . . promised something.
>
> And then television came.

Quoted in Neale, *No Maps for These Territories*, 2000.

7. Jones, *George Lucas*, 2016, 84.

8. Meyrowitz, "Medium Theory," in Crowley and Mitchell, *Communication Theory Today*, 1994, 50–77.

9. Quoted in Ralón, "Interview with Eric McLuhan," 2010.

10. Postman, "The Reformed English Curriculum," in Eurich, *High School 1980*, 1970, 161.

11. See Strate, *Amazing Ourselves to Death*, 2014, esp. chapter 3.

12. Waite, *Mediation and the Communication Matrix*, 2003, 25.

13. Neil Postman explained: "[Marshall McLuhan] used that phrase in a letter he wrote to Claire Booth Luce saying something to the effect that the media ecology of cultures probably need to be managed in a systematic way—and the term struck me as very useful, since *ecology* was used to mean the study of environments and how environments can be healthy and become toxic and so on. I thought that his putting the term *media* in front of ecology suggested in a forceful way that people studying media were not just studying machines and how they work, but the interaction between the structure and form of machines and the human sensorium." Quoted in *Understanding McLuhan*, 1996.

14. Barnes, "Understanding Social Media from the Media Ecological Perspective," in Konijn et al., *Mediated Interpersonal Communication*, 2008, 14–33.

15. Quoted in *Understanding McLuhan*, 1996.

16. Ong, *Interfaces of the Word*, 1977, 305.

17. Ong, *Rhetoric, Romance, and Technology*, 1971, 25.

18. McLuhan, *Counterblast*, 1969, 22.

19. Kittler, *Discourse Networks*, 1990, 245. Nietzsche wrote, "The writing ball is a thing like me: made of iron yet easily twisted on journeys. Patience and tact are required in abundance as well as fine fingers to use us." Nietzsche, "Writing Ball is a thing like me," 1882.

20. Kittler, *Discourse Networks*, 1990, 369.

21. Quoted in Kittler, *Gramophone, Film, Typewriter*, 1999, 200.

22. McLuhan, *Understanding Media*, 1964, 283.

23. As Mark Davidson puts it, "A system, like a work of art, is a pattern rather than a pile. Like a piece of music, it's an arrangement rather than an aggregate. Like a marriage, it's a relationship rather than an encounter." Davidson, *Uncommon Sense*, 1983, 27.

24. Innis, *The Bias of Communication*, 1951, 33.

25. McLuhan, *The Gutenberg Galaxy*, 1962, 24 and throughout.

26. McLuhan, *Understanding Media*, 1964, 160.

27. Though, as Douglas Hofstadter asks parenthetically, "Are we *really* talking to each other if we talk by phone?" Hofstadter, *I Am a Strange Loop*, 2007, 261.

28. Ted Kaczynski arrived at this view. See Kaczynski, *Technological Slavery*, 2008. See also Ellul, *The Technological Society*, 1954.

29. Eno and Mills, *More Dark than Shark*, 1986, 131.

30. See Klein, "I Didn't Want It to Be True," 2022.

31. In the next line he adds, "And this is why Marshall McLuhan is important, more so now than ever, because he saw this coming a long way off, and he saw the reasons for it." Coupland, *Marshall McLuhan*, 2010, 9.

32. I will be using the term *personal media* throughout this book to refer to devices that are used primarily by solitary individuals, including personal stereos, iPods, tablets, and cell phones. See Rasmussen, *Personal Media and Everyday Life*, 2014.

33. Weigend, "The Social Data Revolution(s)," 2009.

34. There's really no such thing as an "average American family." This is according to data from the U.S. Census Bureau: https://web.archive.org/web/20110604174340/http://www .census.gov/compendia/statab/cats/income_expenditures_poverty_wealth.html

35. Carr, *The Shallows*, 2010, 87.

36. Taylor said, "In the past the man has been first; in the future the system must be first." Quoted in Rhodes, *Visions of Technology*, 1999, 45. As for Taylor's reluctant use of the term "scientific management" and the influence of his book, see Postman, *Technopoly*, 1992, 40–55.

37. Gary Marcus, *Kluge*, 2008, 135. Emphasis in original.

38. Murphy, *Last Futures*, 2016, 69. See also Pawley, "Architecture Versus the Movies," 1970.

39. Solnit, *Wanderlust*, 2001, 263.

40. Rosenberger, *Callous Objects*, 2017, 55. Marshall McLuhan wrote, "We are all robots when uncritically involved with our technologies." McLuhan and Fiore, *War and Peace in the Global Village*, 1968, 18.

41. John Gray notes, "Hegel writes somewhere that humanity will only be content when

it lives in a world of its own making." Gray, *Straw Dogs*, 2002, xv. See also Kittler, *Literature, Media, Information Systems*, 1997.

42. Solnit, *River of Shadows*, 2003, 114.

43. Butler, *Erewhon*, 1877, 257.

44. Ibid., 293.

45. Ibid., 259.

46. Eno and Mills, *More Dark than Shark*, 1986, 131.

47. Arthur, *The Nature of Technology*, 2009, 11.

48. Ibid.

49. Kaczynski, 2010, 76–77.

50. Ibid., 77.

51. Though, as John Naisbitt once said, "The two biggest markets in the $8 trillion-dollar-a-year economy of the United States are 1) consumer technology and 2) the escape from consumer technology." Quoted in Boyle, *Authenticity*, 2003, 12.

52. Karl Popper called it "exosomatic evolution," writing, "*Animal evolution* proceeds largely, though not exclusively, by the modifications of organs (or behavior). *Human evolution* proceeds, largely, by developing new organs *outside our bodies or persons*: 'exo-somatically,' as biologists call it, or 'extra-personally'. These new organs are tools, or weapons, or machines, or houses." Popper, *Objective Knowledge*, 1972, 238. Emphasis in source.

53. Quoted in Neale, *William Gibson*, 2000. Dominic Pettman writes, "It is dubious at best to claim that humans can somehow return to a 'natural' state, stripped of all prostheses and artifice." Pettman, *Human Error*, 2011, 7. John Gray adds, "Cities are no more artificial than the hives of bees. The internet is as natural as a spider's web." Gray, *Straw Dogs*, 2002, 16.

54. Lulu, a character in Jennifer Egan's *A Visit from the Goon Squad*, calls it "atavistic purism," which, she explains, "implies the existence of an ethically perfect state, which not only doesn't exist and never existed, but it's usually used to shore up the prejudices of whoever's making the judgments." Egan, *A Visit from the Goon Squad*, 2010, 319.

55. Bradbury, *Fahrenheit 451*, 1953, 84–85. *Texere*, to weave, is the Latin root of both *text* and *texture*. See Alison, *Meander, Spiral, Explode*, 2019, 29.

56. See Gleick, *Chaos: Making a New Science*, 1987. This book changed my life. See "Roy Christopher on Working Through Personal Upheaval and the Mind-Altering Ideas in James Gleick's 'Chaos,'" *Bookshelf Beats*, March 20, 2015, https://medium.com/bookshelf-beats/roy-christopher-on-personal-upheaval-and-the-life-changing-ideas-in-james-gleick-s-chaos-a9531a2aef02.

57. Hayles, *Chaos Bound*, 2018.

CHAPTER 2. AUDIBLE ARRANGEMENTS

Epigraphs are drawn from the following sources: Bloom quote from *The Closing of the American Mind* (New York: Simon & Schuster, 1987), 81. The Nonce quotes from "Mixtapes," track 5 from *World Ultimate* [LP] (New York: American Recordings, 1995). R.I.P., Yusef Muhammad. Home quote from "How I Discovered America," *Info Pool*, No.6, 2002, https://www.stewarthomesociety.org/art/america.htm. Gibson quote from *Distrust That Particular Flavor* (New York: Putnam, 2012), 50. Maughan quote from *Infinite Detail: A Novel* (New York: FSG

Originals, 2019), 156. Toop quote from *Haunted Weather* (San Francisco, Calif.: Serpent's Tail, 2004), 42. Antrim quote from *Elect Mr. Robinson for a Better World* (New York: Vintage, 1993), 129. Chambers quote from *Migrancy, Culture, Identity* (New York: Routledge, 1994), 50. Ullman quote from *Close to the Machine* (San Francisco: City Lights, 1997), 65.

1. Weheliye, *Phonographies*, 2005, 1. Paraphrasing Jon Savage, Mark Fisher wrote, "An album, a single would be a threshold that you could cross that would open up worlds to you. There would be all kinds of references, all kinds of distillations in the cover art." In Butt, Eshun, and Fisher, eds., *Post-Punk*, 2016, 14.

2. Interview with the author, February 6, 2023. The music critic Alex Ross calls the current state of music availability the "Infinite Playlist," adding that it provides "anxiety in place of fulfillment, an addictive cycle of craving and malaise. No sooner has one experience begun than the thought of what else is out there intrudes." Quoted in Gleick, *The Information*, 2011, 409.

3. Valcheva, "Playlistism."

4. Ibid.

5. Gelitz, "You Are What You Like," 2011. Nathan Rabin adds that our "preferences become prejudices." Rabin, *You Don't Know Me*, 2013, 7.

6. Eno, "The Revenge of the Intuitive" 1999, 176. "I like simple instruments," Eno told Deirdre O'Donaghue of KCRW in 1985. "I always have. I've always used very simple synthesizers actually, and I prefer them because I don't particularly care to be faced with limitless possibilities. I prefer a slightly more constrained situation." William Gibson said in 2010, "The last fax I bothered to purchase, which cost virtually nothing, was so loaded with features and options that I've yet to figure out how to send a fax. Fortunately, I only need to send two or three a year, in which case I go to a nearby shop." Quoted in O'Connor, "William Gibson," 2010.

7. Chase, "Philosophy of the Mix," 2005. See also Moore, *Mix Tape*, 2004; Sheffield, *Love Is a Mix Tape*, 2007; Bitner, *Cassette from My Ex*, 2009; Oakes, *Slanted and Enchanted*, 2009, 120–21.

8. Van Dijck, "Remembering Songs Through Telling Stories," in Bijsterveld and van Dijck, *Sound Souvenirs*, 2009, 107–19. Dean Wareham muses, "In the future, when social scientists study the mix-tape phenomenon, they will conclude—in fancy language—that the mix tape was a form of 'speech' particular to the late twentieth century, soon replaced by the 'play list.'" Quoted in Moore, *Mix Tape*, 2004, 28.

9. Moore, *Mix Tape*, 2004, 12.

10. Rombes, *A Cultural History of Punk*, 2009.

11. Quoted in Bromberg, *The Wicked Ways of Malcolm McLaren*, 1989, 232.

12. Quoted in Eshun, *More Brilliant than the Sun*, 2021, 276.

13. Quoted in Bromberg, *The Wicked Ways of Malcolm McLaren*, 1989, 232.

14. Ibid.

15. See Graham, "Malcolm McLaren and the Making of Annabella," in Taylor, *Impresario*, 1988, 59–71.

16. Ball, *I Mix What I Like!*, 2011, 122. In Jacques Attali's political economy of music, mixtapes represent the shift into the last phase of his "networks": sacrifice (performance), representation (notation), repetition (recording), and composition (DIY). Attali, *Noise: The Political Economy of Music*, 1977.

17. Quoted in Bromberg, *The Wicked Ways of Malcolm McLaren*, 1989, 125. According to Creation Records founder Alan McGee, in the mid-1990s McLaren also foresaw the impact of the internet and downloadable MP3s on the music industry. See McGee, *Creation Stories*, 2013. Tellingly, attempts to protect copyright eventually prevented Sony CDs from being played on their own CD players. See F. Rose, "The Civil War Inside Sony," 2003, 100–103, 136–37.

18. Eno, "The Revenge of the Intuitive," 1999, 176.

19. Mark Katz writes, "A phonograph effect indeed, for it is a palpable manifestation of recording's influence. This noise, real or digitally simulated, is now firmly a part of our modern music vocabulary and can be powerfully evocative to listeners. It was long deemed by both the industry and listeners as an unwanted addition to the phonographic experience, but ironically became a valued and meaningful sound when digital technology finally eliminated it. In the age of noiseless digital recordings, this sonic patina prompts nostalgia, transporting listeners to days gone by." Katz, *Capturing Sound*, 2004, 155. Simon Reynolds states, "Like the scratches and surface noise on vinyl, the hiss of tape noise reminds you constantly that this is a recording." Reynolds, *Retromania*, 2011, 350.

20. Eno, *A Year with Swollen Appendices*, 1996, 283. See also Levin, "Indexically Concrète," 1999, 162–69.

21. Shannon Lee Dawdy writes of this patina: "It critiques and it bonds." Dawdy, *Patina*, 2016, 143.

22. For an insightful take on digital files, see Basile, *Tar for Mortar*, 2018, 68–69.

23. Jay Rosen said this about digital images, but neither he nor I were able to recall where. See also Pat Cadigan's story "Rock On," in which music bypasses physical formats altogether and is beamed directly into the brain. Mark Dery writes, "Cadigan's story betrays a mystical humanism that places its faith in the ghost, rather than the machine." Dery, *Escape Velocity*, 1996, 101.

24. Hayles, *How We Became Posthuman*, 1999, 17. See also Tenner, *Our Own Devices*, 2003, xii. As Mary Douglas put it, "Constructing sameness is an essential intellectual activity that goes unobserved." Douglas, *How Institutions Think*, 1986, 60.

25. McGlone, Beck, and Pfiester, "Contamination and Camouflage in Euphemisms," 2006, 261–82.

26. Sterling, "Bruce Sterling on the Art of Text-to-Image Generative AI," 2022.

27. Burningham, *Tilting Cervantes*, 2008.

28. Quoted in Christie, *Gilliam on Gilliam*, 1999, 129.

29. Barry Brummett writes, "Several references within the film make it clear that the characters regard their society as if it were a machine." Brummett, *Rhetoric of Machine Aesthetics*, 1999, 122. Michel de Certeau defines a "culture" as "systems of operational combination." De Certeau, *The Practice of Everyday Life*, 1984, xi.

30. Quoted in Christie, *Gilliam on Gilliam*, 1999, 145. Jeffrey Melton and Eric Sterling describe the world of *Brazil* as "an authoritarian nightmare that in every way discourages human aspiration beyond the desire for comfort and commodities." Melton and Sterling, "The Subversion of Happy Endings in Terry Gilliam's *Brazil*," in Birkenstein, Froula, and Randell, *The Cinema of Terry Gilliam*, 2013, 69.

31. Boyd, "Pastiche and Postmodernism in *Brazil*," 1990, 33–42; Melton and Sterling, "The Subversion of Happy Endings in Terry Gilliam's *Brazil*," in Birkenstein, Froula and Randell, *The Cinema of Terry Gilliam*, 2013. See also the differences between social media (surveillance and distraction) and television (mere distraction) in Lanier, *Ten Arguments for Deleting Your Social Media Accounts*, 2018.

32. See Spufford and Uglow, *Cultural Babbage*, 1996; Babbage, *Passages from the Life*, 1994.

33. Mattern, *Code+Clay . . . Data+Dirt*, 2017.

34. Parikka, *What Is Media Archaeology?*, 2012.

35. Quoted in Danto, *Mark Tansey*, 1992, 135.

36. Lovink, *Dark Fiber*, 2004, 11. Michael Leyton argues that "all cognitive activity proceeds via the recovery of the past through objects in the present." Leyton, *Symmetry, Causality, Mind*, 1992, 2.

37. Bruce Sterling and Richard Kadrey write, "Radio didn't kill newspapers, TV didn't kill radio or movies, video and cable didn't kill broadcast network TV, they just all jostled around seeking a more perfect app." Sterling and Kadrey, "Dead Media Manifesto." James Gleick writes, "The past folds accordion-like into the present." Gleick, *The Information*, 2011, 409.

38. The "archaeology" part of "media archaeology" is largely metaphorical. Mattern, *Code+Clay . . . Data+Dirt*, 2017, xx.

39. See Parikka, *The Anthrobscene*, 2014, esp. 35–51.

40. See Tuhus-Dubrow, *Personal Stereo*, 2017, 34–35. Cal Newport notes, "The headphones of the Sony Walkman and Discman were donned sparingly, for specific times and purposes. The earbuds of the iPod and iPhone are no longer worn exclusively for certain activities but indiscriminately throughout the day." Newport, *Digital Minimalism*, 2019, 100.

41. The stories of the Walkman's origins are from McMahon and Graham, *Introduction to Engineering Materials*, 1992; Du Gay et al., *Doing Cultural Studies*, 1997; Chambers, *Migrancy, Culture, Identity*, 1994; Levy, *The Perfect Thing*, 2006; and Hagood, *Hush*, 2019.

42. Quoted in Headlam, "Origins," 1999.

43. Ibid.

44. Du Gay, et al., *Doing Cultural Studies*, 1997, 23.

45. Bull, *Sounding Out the City*, 2000, 96. William Gibson agrees, saying, "I remember I bought the first Walkman I ever saw. [. . .] The experience of taking the music of your choice and being able to move it through the environment of your choice had just never been available. It felt weirdly subversive. I could walk through rush-hour crowds listening to Joy Division at skull-shattering volumes [laughs], and no one knew. Like you were having this completely different experience that was completely altering the way it all looked and no one knew." Quoted in Kodwo Eshun, "William Gibson: The Co-Evolution of Humans and Machines," in Christopher, *Follow for Now, Vol. 2*, 2021, 276. See also Prescott-Steed, "Frostbite on My Feet," 2013, 45–68.

46. Quoted in Bull, *Sounding Out the City*, 2000, 96. As Joshua Meyrowitz writes of the Walkman, "Through such media, what is happening almost everywhere can be happening wherever we are. Yet when we are everywhere, we are also no place in particular." Meyrowitz, *No Sense of Place*, 1985, 125.

47. Bull, *Sounding Out the City*, 2000, 96. Brandon LaBelle adds, "The wired-up walker enacts a sort of ghosting of the sidewalk—we may never know what sonic matter is floating through the ear of the iPod user, whose step occupies the vague threshold between zombism and activism." LaBelle, *Acoustic Territories*, 2010, 98.

48. David Toop compares our use of the Walkman to the ambient chamber music enjoyed by royalty, saying that we've all become monarchs. See Toop, *Ocean of Sound*, 1995, 271. See also Chow, "Listening Otherwise, Music Miniaturized," 1990, 129–48; Sommer, *Personal Space*, 1969, 40; Tuhus-Dubrow, *Personal Stereo*, 2017, 3.

49. Jansen, "Tape Cassettes and Former Selves," in Bijsterveld and van Dijck, *Sound Souvenirs*, 2009, 43–54; Boym, *The Future of Nostalgia*, 2001. The Welsh term for this longing is "hiraeth," Walter Chaw writes, "nostalgia for something you never experienced—that maybe never existed, but is truer than memory." See Chaw, *A Walter Hill Film*, 2023, 206.

50. Grainge, *Monochrome Memories*, 2002.

51. Huyssen, *Present Pasts*, 2003, 1.

52. Chambers, *Migrancy, Culture, Identity*, 1994, 51.

53. Guattari, *Machinic Eros*, 2015, 97.

54. Du Gay et al., *Doing Cultural Studies*, 1997, 16–17.

55. Meyrowitz, *No Sense of Place*, 1985, 90.

56. Lanois, *Soul Mining*, 2010, 172.

57. Ibid., 173.

58. Bull, *Sound Moves*, 2007, 7. See also Hagood, *Hush*, 2019, 12.

59. Hagood, *Hush*, 2019, 218.

60. Wilson Rothman writes in an essay celebrating the thirty-year anniversary of the Walkman, "The other big advantage of cassettes, of course, [was] that they were recordable. You'd buy blank 90-minute cassettes (chrome high bias, if you were an audio nut) and tape one album on each side. (Since most records were shorter than 45 minutes, you'd grab a song or two from another album to avoid a long dead spot before the tape reversed.) And you'd borrow albums from friends and tape your own. You could also tape from other cassettes, but the quality degraded each time you made a copy made from a copy. It was like an organic form of DRM. Everybody had a box with hand-labeled cassettes and before you went on a car trip, you'd dig in the box to find the tunes that would soundtrack your journey." Rothman, "The Blank Generation," 2009.

61. W. Gibson, "The Recombinant City: A Foreword," in Delany, *Dhalgren*, 1996, xii.

62. Weheliye, *Phonographies*, 2005, 135. See also Hagood, *Hush*, 2019, 218; Du Gay et al., *Doing Cultural Studies*, 1997, 59.

63. Latour, *Pandora's Hope*, 1999; Rosenberg, *Inside the Black Box*, 1982.

64. Chambers, *Migrancy, Culture, Identity*, 1994, 49–50. William Gibson adds, "The Walkman changed the way we understand cities." W. Gibson, *Distrust That Particular Flavor*, 2012, 13. Adam Krims adds, "The mutual encoding of music genre and urban geography also works through musical poetics, though never purely or in an unmediated way." Krims, *Music and Urban Geography*, 2007, 17.

65. Quoted in Block and Glasmeier, *Broken Music*, 1989, 73. See also Grubbs, *Records Ruin*

the Landscape, 2014. Cage had, of course, already used records to make compositions with his series of *Imaginary Landscapes* (1939–1952). See Prendergast, *The Ambient Century*, 2000, 44–49.

66. Quoted in "Christian Marclay" (interview), 1993, 26.

67. Quoted in Szendy, "Christian Marclay on the Phone," in *RE:Play*, 2007, 89.

68. Quoted in Taylor, Sharp, and Higgs, *pressPLAY*, 2005, 10.

69. Marclay, *Christian Marclay*, 2003, 89.

70. Quoted in Criqui, *On and by Christian Marclay*, 2014, 42. Marclay tells Ben Neill, "I'm not very high tech. Most of my work has been very low tech. I was never interested in all that hi-fi sound quality. I wanted to exploit the drawbacks of the technology." Neill, "Christian Marclay," 2003, 49.

71. Shapiro, "Deck Wreckers," in Young, *Undercurrents*, 2002, 163–76.

72. Quoted in Khazam, "Jumpcut Jockey," in Criqui, *On and by Christian Marclay*, 2014, 31. I call the idea of recorded memories a broken metaphor, though Freud did once describe someone with excellent recall as having a "phonographic memory." Freud, "Fragment of an Analysis of a Case of Hysteria," 1905, 10.

73. Quoted in Ferguson, "Never the Same Twice," in Criqui, *On and by Christian Marclay*, 2014, 76.

74. Greil Marcus, *Lipstick Traces*, 1989, 2. Marcus writes, "Is history simply a matter of events that leave behind those things that can be weighed and measured—new institutions, new maps, new rulers, new winners and losers—or is it also the result of moments that seem to leave nothing behind, nothing but the mystery of spectral connections between people long separated by place and time, but somehow speaking the same language?" Ibid., 4. Marclay tells Douglas Kahn, "When I first came to the United States, it was a common sight to see broken records on the street." Quoted in Kahn, "Christian Marclay's Early Years," 2003, 19.

75. Marclay, *Record Without a Cover*, 1985. See also Dworkin, *No Medium*, 2013, 148–49; Block and Glasmeier, *Broken Music*, 1989, 177.

76. As the character Harrisch puts it in K. W. Jeter's *Noir*, "All that cute blather people talked about a while back, about how the future would be nothing but little bits of information being zipped back and forth, the whole world on-line and freed of the constraints of gross materiality—that didn't come to pass. Atoms endure." Jeter, *Noir*, 1998, 140.

77. Hegarty, *Noise/Music*, 2008, 182.

78. Shapiro, "Deck Wreckers," 2002.

CHAPTER 3. ALGORITHM NATION

Epigraphs are drawn from the following sources: Robinson quote from *2312: A Novel* (New York: Orbit, 2012), 285. Gibson quote from *Virtual Light* (New York: Bantam, 1993), 284. Easterling quote from *Medium Design* (N.p.: Strelka Press, 2018), 1. Rushkoff quote from *Team Human* (New York: W. W. Norton, 2019), 158. Gray quote from *Straw Dogs: Thoughts on Humans and Other Animals* (New York: Farrar, Straus and Giroux, 2002), 15. Attali quote from *Noise: The Political Economy of Music* (Minneapolis: University of Minnesota Press, 1977), 20.

1. See chapter 10 ("Sampling, Tinkering and the Glitch") in Rhodes and Westwood, *Critical Representations of Work and Organization*, 2008, 172–97, esp. 177.

2. Eno, foreword to Prendergast, *The Ambient Century*, 2000, xi–xii.

3. Saval, "Wall of Sound," 2011.

4. Quoted in Licht, "CBGB as Imaginary Landscape," in Ferguson, *Cristian Marclay*, 2004, 97.

5. Sterne, *The Audible Past*, 2003, 293.

6. Ibid., 292. Curiously, motion picture pioneer Eadweard Muybridge contacted Edison on February 27, 1888, to suggest that they combine efforts and reproduce "in the presence of an audience, visible actions and audible words." See Muybridge, *Animals in Motion*, 1957, 15.

7. Andreas Huyssen muses, "Could it be that the surfeit of memory in this media-saturated culture creates such an overload that the memory system itself is in constant danger of imploding, thus triggering fear of forgetting?" Huyssen, *Present Pasts*, 2003, 17. David Toop adds, "Without wishing to minimize the value of archiving or the very real problems posed by multiple formats, rapid technological change, and obsolescence, the focus on resource seems to confirm our obsession with preserving the past rather than confronting the present." Toop, *Haunted Weather*, 2004, 73.

8. Norman, *Turn Signals Are the Facial Expressions of Automobiles*, 1992, 1.

9. Ibid., 2.

10. Gitelman, "Unexpected Pleasures," in Eglash et al., *Appropriating Technology*, 2004, 331–44. Edison had been trying to synchronize the phonograph with the kinetoscope, but two problems remained until the 1920s: synchronization and amplification. See Chanan, *Repeated Takes*, 1995.

11. Quoted in Novak, "Musicians Wage War Against Evil Robots," 2012. In 1970 Aldous Huxley wrote, "Recreation is provided ready-made by enormous joint-stock companies. . . . In the days before machinery men and women who wanted to amuse themselves were compelled, in their humble way, to become artists. Now they sit and permit professionals to entertain them by the aid of machinery." Huxley, *America and the Future*, 1970, 11.

12. Hegarty and Halliwell, *Beyond and Before*, 2011. William Gibson writes, "I took Punk to be the detonation of some slow-fused projectile buried deep in society's flank a decade earlier." From Gibson's old website, quoted in Rombes, *A Cultural History of Punk*, 2009, 108.

13. Reynolds was talking specifically about the no wave movement. See Reynolds, *Rip It Up and Start Again*, 2005, 157. About punk, Will Sargeant writes, "This extremely powerful movement was designed to purge the old and sweep the decks ready for what was to come next." Sargeant, *Bunnyman*, 2021, 173.

14. Quoted in Berden, "Robert Smith," 1989. Brian Eno was commissioned by the Sony Corporation in 1984 to record a piece for exclusive release on compact disc. His 1985 record *Thursday Afternoon* was the result. The CD was the only format that would accommodate the uninterrupted sixty-one-minute run time of the composition. See Leung, "Figure and Ground, Image and Sound," 2016, 91–105; Bofop, "*Thursday Afternoon*," 1985.

15. See C. Moore, "Works and Recordings," in Talbot, *The Musical Work*, 2000, 88–109.

16. Bull, "Investigating the Culture of Mobile Listening," in O'Hara and Brown, *Consuming Music Together*, 2006, 132.

17. Katz, *Capturing Sound*, 2004, 171.

18. Email to the author, April 17, 2009.

19. RIAA, "U.S. Recorded Music Revenues by Format, 1973 to 2020."

20. Richter, "The Rise and Fall of the Compact Disc," 2022. Brian Eno employs an apt metaphor for their dilemma: whale blubber as an energy source:

> I think records were just a little bubble through time and those who made a living from them for a while were lucky. There is no reason why anyone should have made so much money from selling records except that everything was right for this period of time. I always knew it would run out sooner or later. . . The record age was just a blip. It was a bit like if you had a source of whale blubber in the 1840s, and it could be used as fuel. Before gas came along, if you traded in whale blubber, you were the richest man on Earth. Then gas came along, and you'd be stuck with your whale blubber. Sorry mate—history's moving along.

Quoted in Morley, "On Gospel, Abba, and the Death of the Record," 2010.

21. Email to the author, April 17, 2009.

22. Email to the author, April 18, 2009.

23. Email to the author, April 18, 2009.

24. Quoted in Ganz and Rose, "The MP3: A History of Innovation and Betrayal," 2011.

25. The first portable MP3-player was the Rio PMP300, manufactured by a small Korean company called Diamond Multimedia. As Michael Denning writes, "Unlike modern novels, paintings, theater, or even film, which 'represented' the modern 'masses,' these discs circulated the voices of those masses. And unlike our postmodern moment when high and low, experimental and pop, mix indiscriminately on the same playlist, modernism was a time of discrimination, when deference and defiance met." Denning, *Noise Uprising*, 2015, 4. See also Barfe, *Where Have All the Good Times Gone?*, 2004, esp. 324–44.

26. Shorter songs also rack up larger play counts. See Pearce, "Considering the Rise of the Super Short Rap Song," 2018. Harold Innis thought the same about long books for rental libraries: the longer the book, the longer the rental. Christian, *The Idea File of Harold Adams Innis*, 1980, 128.

27. The 78-rpm album, with only three minutes per side, was responsible for the three-minute pop song. Bofop, "*Thursday Afternoon*," 1985.

28. Quoted in Coscarelli, "15 Songs in 15 Minutes," 2018.

29. See McLaren, *Musical Paintings*, 2009, 6.

30. Friedrich Kittler wrote, "The snow that helped trackers was an accident; Edison's tinfoil roll or Francis Galton's fingerprint archive were purposefully prepared recording surfaces for data that could be neither stored nor evaluated without machines." Kittler, *Discourse Networks 1800/1900*, 1990, 237.

31. Nelson, *Literary Machines*, 1987, 3/25.

32. Quoted in Sinker, *We Owe You Nothing*, 2001, 147.

33. Nelson, *Literary Machines*, 1987, 3/24.

34. Quoted in Gordon, "The Evolution of Steve Albini," 2023.

35. Nelson, *Literary Machines*, 3/25. Nicholas Carr writes, "To remain vital, culture must be renewed in the minds of the members of every generation. Outsource memory, and culture

withers." Carr, *The Shallows*, 2010, 197. Jaron Lanier adds, "The culture of computation has always been a cult of youth, which doesn't prioritize remembering." See Jaron Lanier's introduction to the 2012 version of Ullman, *Close to the Machine*, 2012, xii.

36. See Rosen, "The Day the Music Burned," 2019.

37. Kleinman, "MySpace Admits Losing 12 Years' Worth," 2019.

38. Ibid.

39. Quoted in ibid. Months of accumulated files on Google Drive recently suffered the same fate. See Toulas, "Google Drive Users Angry over Losing Months of Stored Data," 2023. In the digital era, this is only likely to get worse. See Linge, "The Archaeology of Communications' Digital Age," 2013.

40. Nelson, *Literary Machines*, 3/25.

41. As Albert Gaines says in Omar El Akkad's novel *American War*, "The first thing they try to take from you is your history." Akkad, *American War*, 2017, 150.

42. See Weick and Roberts, "Collective Mind in Organizations," 1993, 357–81. Daniel Lanois writes, "*Remembering* is just another word for choosing." Lanois, *Soul Mining*, 2010, 13.

43. Brand, *The Clock of the Long Now*, 1999.

44. See ibid., esp. 81–92.

45. Urban, *Metaculture*, 2001, 42. See also ibid., 73. These physical manifestations are also the earliest signs of consciousness. See Corson, "Speed and Technicity (a Derridean Exploration)," 2000.

46. Urban, *Metaculture*, 2001, 53–55.

47. Streaming services don't seem to be equipped to handle any of this any better. See Joe Pinsker, "What Will Happen to My Music Library?" As Tim Maughan imagines it after the fall of the internet, "Decades of history, long lost elsewhere, [are] spoken on vinyl in the machine language." Maughan, *Infinite Detail*, 2019, 162.

48. See Sinker, *We Owe You Nothing*, 2001.

49. See Greenwald, *Nothing Feels Good*, 2003, 147–48.

50. Quoted in Azerrad, *Our Band Could Be Your Life*, 2001, 399.

51. MacKaye explains,

> [The Fugazi song] "Merchandise" was a response to the overarching emphasis on merchandise at shows in the mid-1980s. At these shows there was so much energy going into these bands selling stuff that in my mind the whole point of the music became trivialized. They were practicing this standard capitalist form of drawing in clientele. You draw in an audience, and they become your clientele. It was like the old snake oil salesman. They would travel through the frontier, and they'd have a caravan of musicians, acrobats or whatever. They'd set up in the middle of a town and do a show. People would gather and in between the acts the 'doctor' would come out with his various tinctures and oils which were all alcohol, essentially. This is exactly the same story in bars today. The idea that you have to see a band in a bar is so odious. It's actually that same practice, music being used as a shill to sell other products.

Quoted in Azerrad, 2001, 397.

52. Clayton, *Uproot*, 2016, 136. Paul Hegarty writes of Minor Threat, "The band's music acts

as a conduit for a community to come into being; the songs are its constitution and the lyrics, the defining philosophy." Hegarty, *Rumour and Radiation*, 2015, 85.

53. Fugazi Live Series, Discord Records, https://www.dischord.com/fugazi_live_series.

54. Quoted in Kurland, "Getting Deep with Ian MacKaye," 2015.

55. Interview with the author, February 6, 2023.

56. Ibid.

57. Quoted in Nedorostek and Pappalardo, *Radio Silence*, 2008, n.p.

58. Quoted in Hyden, "How Radiohead's 'Kid A' Kicked Off the Streaming Revolution," 2015. Brackets in original.

59. Ibid.

60. DeSantis, "Radiohead's Digital Album Sales, Visualized," 2016.

61. Ibid.

62. I'm borrowing the concept of *novelty* from Terence McKenna's Timewave and the idea of *nodal points* from William Gibson's 1996 novel *Idoru*. The former is a computer-generated timeline based on chaos theory and the I Ching, in which the peaks represent increased human novelty (e.g., artistic innovation and scientific discovery). The latter is a sort of subconscious pattern recognition where certain seemingly mundane data converge into sharp points of interest. Influential and classic cultural artifacts like records are excellent examples of both. See McKenna, *The Archaic Revival*, 1992; and W. Gibson, *Idoru*, 1996.

63. The Buzzcocks, the Smiths, the Fall, Echo and the Bunnymen, and Joy Division/New Order all came out of that gig. It also launched the Manchester music scene, Factory Records, and the Hacienda nightclub. See Nolan, *I Swear I Was There*, 2016.

64. Fisher, *The Ghosts of My Life*, 2014, 50.

65. Quoted in Gross, "Gang of Four," 2000.

66. Dettmar, *33 1/3: Entertainment!*, 2014, 36.

67. See Butt, Eshun, and Fisher, *Post-Punk*, 2016, 18, 122, and throughout; Dettmar, *33 1/3: Entertainment!*, 2014; and Reynolds, *Rip It Up and Start Again*, 2005.

68. Quoted in Singer, "Hotseat," 2014.

69. Quoted in Christopher, "Dave Allen: Every Force Evolves a Form," in *Follow for Now, Vol. 2*, 2021, 117.

70. Quoted in Singer, "Hotseat," 2014.

71. Rita Raley writes, "To articulate tactical media in terms of performance rather than as static art object emphasizes viewer experience and engagement. . . . To conceive of tactical media in terms of performance is to point to a fluidity of its actants, to emphasize its ephemerality, and to shift the weight of emphasis to the audience, which does not simply complete the signifying field of the work but records a memory of the performance." Raley, *Tactical Media*, 2009, 12.

72. Quoted in Christopher, "El-P: Wake Up. Time to Die," in *Follow for Now, Vol. 2*, 2021, 194.

73. Ibid.

74. Killer Mike, "Killer Mike Baltimore Op-Ed," 2015.

75. Paine, "Killer Mike and El-P 'Run The Jewels,'" 2013.

76. Van der Doelen, *Kill Your Masters*, 2024, 85–86.

77. Quoted in Open Mike Eagle, "What Had Happened Was with El-P," 2021.

78. Van der Doelen, *Kill Your Masters*, 2024, 85.

CHAPTER 4. TIME OF THE SIGNS

Part 2 epigraphs are drawn from the following sources: Tanaka quoted in Félix Guattari, *Machinic Eros: Writings on Japan* (Minneapolis, Minn.: Univocal, 2015), 46. Bourdieu quote from *The Logic of Practice* (Stanford, Calif.: Stanford University Press, 1990), 228. Porphyrus quoted in Gaston Bachelard, *The Poetics of Space* (Boston: Beacon Press, 1964), 223.

Chapter 4 epigraphs are drawn from the following sources: Gibson quote from *All Tomorrow's Parties* (New York: Putnam, 1999), 174–75; Burgess quote from *A Clockwork Orange* (London: William Heinemann, 1962), 114. Keyes quote from *Flowers for Algernon* (New York: Harcourt, Brace & World, 1966), 207. Miller quoted in M. Keith Booker, *Comics through Time: A History of Icons, Idols, and Ideas*, vol. 3 (Westport, Conn.: Greenwood Publishing, 2014), 1519. Kay quoted in Kevin Kelly, *What Technology Wants* (New York: Viking, 2010), 235. Negarestani quote from *Cyclonopedia: Complicity with Anonymous Materials* (Melbourne, Australia: re.press, 2008), xvii. McCarthy quote from *The Passenger* (New York: Knopf, 2022), 190. Thompson quote from *Rosewater* (Lexington, Ky.: Apex Book Company, 2016), 98.

1. Deverson and Hamblett, *Generation X*, 1964, 109.

2. Quoted in "Heap of Hype Heralds Sigue Arrival," 1986.

3. Quoted in Leigh, "I just kept cool, you know," 2001.

4. Needs, "Sigue Sigue Sputnik," 2020, 128.

5. Dery, *Escape Velocity*, 1996, 76.

6. See *Reality Bites*, 2024.

7. Gordinier, *X Saves the World*, 2008. Gordinier calls this mix of "the traditional and the transgressive" a Gen X trademark (36), "the desire for change and profound doubts about how to achieve it" (15). See also Purdy, *For Common Things*, 1999, 96.

8. See Haynsworth, "'Alternative' Music and the Oppositional Potential," in Ulrich and Harris, *GenXegesis*, 2003, 41–58; Brooks, "In the 90s, We Worried about Nirvana," 2023.

9. Mark Fisher wrote in 2009, "'Alternative' and 'independent' don't designate something outside mainstream culture; rather, they are styles, in fact *the* dominant styles, within the mainstream." Fisher, *Capitalist Realism*, 2009, 9.

10. Germs, "What We Do Is Secret!," 1979. See also Lingel, *Digital Countercultures and the Struggle*, 2017, esp. 71–97.

11. Coupland, "Douglas Coupland on Generation X at 30," 2021. Gene Youngblood wrote, "To understand this generation gap, we must realize that the melancholy of the new nostalgia arises not out of sentimental remembrance of things past, but from an awareness of radical evolution in the living present." Youngblood, *Expanded Cinema*, 1970, 143.

12. William Strauss and Neil Howe define a generation as a group of people "whose common location in history lends them a collective persona," and within their theoretical framework Generation X is "the Nomad." See Strauss and Howe, *The Fourth Turning*, 1997.

13. "Generational Insights and the Speed of Change," 2022. As Heraclitus wrote in "Fragment 88," "Thirty, therefore, names the moon of generation." Heraclitus, *Fragments*, 2001, 55.

14. Brand, *The Clock of the Long Now*, 1999.

15. Miroshnichenko, *Human as Media*, 2020, 43. See also Rushkoff, *Playing the Future*, 1996; Pettman, *Look at the Bunny*, 2013.

16. As Sherry Turkle points out, "To generations that grew up using their phones to text and message, *these studies may be describing losses they don't feel.*" Turkle, *Reclaiming Conversation*, 2015, 13. Emphasis in original.

17. Harris, *The End of Absence*, 2014, 15.

18. Ibid. See also Power, "Why Gen X Failed," 2022.

19. Cunningham, "Sales of Vinyl Albums Overtake CDs," 2023.

20. Toop, "Christian Marclay," 2011, 43.

21. Dominic Pettman points out that all of our media are incubators for new totems, connecting us to media of the past. See Pettman, *Look at the Bunny*, 2013.

22. Greenwald, *Nothing Feels Good*, 2003, 56.

23. Urban, *Metaculture*, 2001, 41.

24. Fogarty, "Each One Teach One," in Bennett and Hodkinson, *Ageing and Youth Cultures*, 2012, 55.

25. Quoted in Kelly, "Mr. Big Trend," 1994, 115.

26. Van Gennep, *The Rites of Passage*, 1960; Douglas, *Purity and Danger*, 1966.

27. Thomas de Zengotita writes of digitally zombified youth, "It was if they were somnambulating, hypnotized, into some newborn zone of being where hallowed custom and bizarre context were so surreally fused that the whole tableau seemed poised to shimmer off into the ether at any moment." De Zengotita, *Mediated*, 2005, 155.

28. Fisher, *K-Punk*, 2018, 57.

29. Du Gay et al., *Doing Cultural Studies*, 1997, 85; Rheingold, *Net Smart*, 2012.

30. Watkins, *The Young and the Digital*, 2009. In trying to synchronize sound with moving pictures, amplification remained an a problem until the 1920s because of the large spaces where people gathered to see movies. See Chanan, *Repeated Takes*, 1995, 71.

31. McLuhan, *Understanding Media*, 1964; Turkle, *Alone Together*, 2011, 152.

32. Bull, "Automobility and the Power of Sound," 243–59.

33. A black box is "a system that is not technically understood or accessed" by its user. Parikka, *A Geology of Media*, 2015, 148–49. As Kristen Gallerneaux, the curator of technology collections at the Henry Ford Museum, writes, "The objects under my curatorial care are essentially a huge collection of Latourian black boxes. They exist as physical proof that the more seamless and successful a technology is, the more mystifying and opaque its inner functions become to the everyday user." Gallerneaux, *High Static Dead Lines*, 2018, 7. See also Latour, *Pandora's Hope*, 1999; Rosenberg, *Inside the Black Box*, 1982.

34. Braidotti, *Nomadic Subjects*, 1994, 17.

35. McLuhan, *Understanding Media*, 1964, 391. He added, "Electronic man is no less a nomad than his Paleolithic ancestors." Ibid., 283.

36. For further discussion of this fallacy, see Postman, *Technopoly*, 1992, 56–70.

37. Eisenstein, "Montage of Attractions," 1974, 77–85, 78. Gerald Raunig adds, "They had to expand the machine concept from the body-machines of the actors and the machine constructions on the stage to the social machine, which stretched beyond the protagonists on the

stage to a diffuse and illimitable assemblage: it was the viewers that should finally be inflamed by the trained elastic actor-machine and the constructive apparatuses." Raunig, *A Thousand Machines*, 2010, 48–49.

38. Augustine, *Confessions*, 1961, 37. This is evidence of Saint Augustine's "two wills"—one to serve God and one to serve himself—that unite his concerns with those of other philosophers. See Critchley, *ABC of Impossibility*, 2015, 45–46.

39. Ibid., *Confessions*, 1961, 39. As John Gray writes, "It is only in sermons or on the stage that human beings are enabled by extremes of suffering." Gray, *Straw Dogs*, 2002, 99.

40. Waite, *Mediation and the Communication Matrix*, 2003, 88–91. 98–99.

41. Buchloh, "Andy Warhol's One-Dimensional Art," in Michelson ed., *Andy Warhol*, 2001, 28.

42. Sirius, "The Importance of Being Andy," 1989, 85.

43. The 1991 Momus essay was published by the Swedish fanzine *Grimsby Fishmarket* in 1992 and in the daily paper *Svenske Dagblatt* in 1994. Momus, "Pop Stars? Nein Danke!," 1991.

44. Weinberger, *Small Pieces Loosely Joined*, 2002, 103–4.

45. C. Anderson, *The Long Tail*, 2006. As Rick Moody puts it, "There are only niche markets." Moody, foreword to Woodworth and Grossan, *How to Write About Music*, 2015, xi. See also Jacobs, *The Economy of Cities*, 1969, 146–47.

46. Shadyac, *Bruce Almighty*.

47. Joshua Meyrowitz writes, "Many jokes, phrases, expressions, and events heard and seen on television provide a common set of 'experience' for people across the land." Meyrowitz, *No Sense of Place*, 1985, 145.

48. Quoted in Molinaro, McLuhan, and Toye, *Letters of Marshall McLuhan*, 1987, 473–74. Italics in original. How much one relies on the ground to interpret the figure is what the cognitive psychologist Herman Witkin called *field dependence*. Witkin and Goodenough, *Cognitive Styles, Essence and Origins*, 1981.

49. As cognitive scientist Gary Marcus notes, "The thing about context is that it is always with us. . . . Just about every time we remember *anything*, context looms in the background." Gary Marcus, *Kluge*, 2008, 24.

50. Riley, *After the Mass-Age*, 2017, 26.

51. Ibid., 37. See also Rushkoff, *Team Human*, 2019, 63.

52. We have become what Jaron Lanier calls "the commenting class." See Lanier, *Dawn of the New Everything*, 2017, 329.

53. Eno, *A Year with Swollen Appendices*, 1996, 328.

54. Eno, "Perfume, Defense, and David Bowie's Wedding," in Scoates, *Brian Eno*, 2013, 227.

55. Granovetter, "The Strength of Weak Ties," 1973.

56. Quoted in Constantino and Nuñez, "Networks, Weak Ties, and Thresholds," 2019.

57. Robertson, *None of This Is Normal*, 2018, 116.

58. Wenger, *Communities of Practice*, 1999.

59. Star and Griesemer, "Institutional Ecology, 'Translations' and Boundary Objects," 1989.

60. Quoted in BBVA Foundation, "Interview with Mark Granovetter," 2022.

61. Quoted in Ross, "How Björk Broke the Sound Barrier," 2015.

62. C. Anderson, *The Long Tail*, 2006, 2.

63. Andy Greenwald writes, "It is simply no longer possible for one pop group to dominate the culture to the degree that the Beatles did in the '60s, or Michael Jackson did in the '80s, maybe not even like Nirvana did in the '90s—there's far too much culture." Greenwald, *Nothing Feels Good*, 2003, 71. See also ibid., 142.

64. Simon Reynolds describes it as "a hipster stock market based around trading in pasts, not futures." Reynolds, *Retromania*, 2011, 419.

65. Lanier, *Who Owns the Future?*, 2013, 11.

66. See C. Anderson, *The Long Tail*, 2006, 10, 22–23. The 1990s were what my friend Dave Allen called at the time "the clutter of pop" in his zine *The Clutter of Pop.*

67. Beachy, *The Most Fun Thing*, 2021.

68. See Kugelberg's foreword to Craig, *I'll Kick You in the Head with My Energy Legs*, 2012.

69. Bakker, *Neuropath*, 2009, 279.

70. The linguists Cliff Goddard and Anna Wierzbicka describe cultural scripts as "common sayings and proverbs, frequent collocations, conversational routines and varieties of formulaic or semi-formulaic speech, discourse particles and interjections, and terms of address and reference—all highly 'interactional' aspects of language." Goddard and Wierzbicka, "Cultural Scripts," 2004, 154.

71. Bret Easton Ellis puts it in terms of a monolithic empire: "If Empire was about the heroic American figure—solid, rooted in tradition, tactile, and analog—then post-Empire was about people who were understood to be ephemeral right away; digital disposability doesn't concern them—they're rooted in traditions created by social media, which is solely about exhibition and surface." Ellis, *White*, 2019, 211.

72. Sconce, *Haunted Media*, 2000, 18.

73. Chambers, *Migrancy, Culture, Identity*, 1994, 64.

74. Gerald Raunig writes, "In so many situations it appears as though machines were not penetrating into human beings as much as humans are being drawn 'into the machine.'" Raunig, *Dividuum*, 2016, 111.

75. Eno adds, "Then in a certain moment, I lose control and at last I am part of the machinery." Eno and Mills, *More Dark than Shark*, 1986, 131. Peter Gabriel, in contrast to Eno, says, "I've had minimal drug experiences because of fear. I would be afraid of losing control. I can trust machines, yet I can't trust pills. . . A machine you can always switch off, or get out of . . . whereas when a pill gets ahold of your metabolism, you have to ride through." Quoted in Easlea, *Without Frontiers*, 2014, 152.

76. Eno, "The Revenge of the Intuitive," 1999. Eno adds, "This transfer is not paying off."

77. Arnold van Gennep identifies three categories of ceremonial passage: *rites of separation, transition rites*, and *rites of incorporation*. Those phases run parallel to the sections of this book. Van Gennep, *The Rites of Passage*, 1960, 11.

78. Heim, *The Metaphysics of Virtual Reality*, 1993, 97.

79. As Jeffrey Sconce writes, "Where there was once 'depth' and 'affect,' there is now only 'surface.' Where there was once 'meaning,' 'history,' and a sold realm of 'signifieds,' there is now only a haunted landscape of vacant and shifting signifiers." Sconce, *Haunted Media*, 2000, 171.

CHAPTER 5. THE SURFACE INDUSTRY

Epigraphs are drawn from the following sources: Baudrillard quote from *Simulacra and Simulations* (Stanford, Calif.: Stanford University Press, 1988), 45. Massumi quote from *A User's Guide to Capitalism and Schizophrenia* (Cambridge, Mass.: MIT Press, 1992), 134. Delany quote from *Dhalgren* (New York: Bantam, 1975), 149–150. Vacheron quote from "Venice Arena," in Francesco Ragazzi, *Palm Angels* (New York: Rizzoli, 2014), CI. Bull quote from "The Audio-Visual iPod," in *The Sound Studies Reader*, ed. Jonathan Sterne (New York: Routledge, 2012), 200. McLuhan quote from *Understanding Media: The Extensions of Man* (New York: Houghton Mifflin, 1984), 98. Kaufman quote from Screenwriters' Lecture, BAFTA, September 30, 2011. Ward quote from the introduction to *The Certeau Reader*, ed. Graham Ward (Oxford: Blackwell, 2000), 7. Anderson quoted in Bill Black, "A Voice from Outer Space," *Sounds*, June 7, 1986, 40.

1. Quoted in Sinker, *We Owe You Nothing*, 2001, 13. Pharrell Williams uses the same metaphor, writing, "The love of the board supplied us with a pair of lenses. I still wear them today. I couldn't take them off—even if I tried. Ask anybody who has ever been on a skateboard, they'll tell you the same thing." P. Williams, foreword to Ragazzi, *Palm Angels*, 2014, iv–v. Pro skateboarder Mike Vallely puts it this way, "We lived in a dead-end town with nowhere to go, and then when skateboarding came along—especially street skating—it made our town bearable. It made our town livable because it blew it wide open. . . It was like the possibilities were suddenly endless. . . . Suddenly it was like, this place ain't so bad because there's a curb and there's a bench and there's some stairs and there's a wall. Everything was redefined. These weren't things that confined and defined our lives, they were things we were now defining." In Peralta, *Bones Brigade*, 2012.

2. Interview with the author, February 6, 2023.

3. See Hawk, "From Carving to Flying," in Zaki, *California Concrete*, 2019, 6–9.

4. Zarka, *On a Day with No Waves*, 2011, 111.

5. Johan Kugelberg writes, "Unlike other 'street artists,' for instance the graffiti painter, the skateboarder's audience of peers is secondary to their own self-evaluation. It is a mirror that rarely flatters. Here geographical mastery is achieved through performance; they are aesthetes as much as they are athletes." Kugelberg, foreword to Craig, *I'll Kick You in the Head with My Energy Legs*, 2012,

6. Brian Schrank, "Research Statement," http://www.brianschrank.com/Research_Statement_Schrank.pdf.

7. J. Gibson, *The Senses Considered as Perceptual Systems*, 1966. Gibson, who is credited with innovating the concept of *affordances*, distinguishes between them and an object's intrinsic properties. See J. Gibson, *The Ecological Approach to Visual Perception*, 1979. See also Norman, *Turn Signals Are the Facial Expressions of Automobiles*, 1992, 19.

8. Jessa Lingel writes, "No technology is single use. Whether from hapless accident or deliberate tinkering, technology is always subject to forms of appropriation and play, misuse and reuse, often in ways that are unintended and unimagined by designers and inventors." Lingel, *Digital Countercultures and the Struggle for Community*, 2017, 1.

9. Houston, "Xerox and Roll," 2023.

10. Ibid.

11. Benjamin, "The Work of Art in the Age of Mechanical Reproduction," 1968, 221.

12. Eichhorn, *Adjusted Margin*, 2016, 25–26. As Friedrich Kittler put it, "Technologically possible manipulations determine what can in fact become a discourse." Kittler, *Discourse Networks*, 1990, 232. See also Oakes, *Slanted and Enchanted*, 2009, esp. 79–96. As McLuhan wrote in 1970, "Gutenberg made everyone a reader, Xerox made everyone a publisher." McLuhan and McLuhan, *Laws of Media*, 1988, 145. As Todd Gitlin writes, "Poetry and song migrated across Europe hand to hand, mouth to ear to mouth. Broadsheets circulated. From the second half of the fifteenth century on, Gutenberg's movable type made possible mass-printed Bibles and a flood of instructional as well as scurrilous literature. Even where literacy was rare, books were regularly read aloud." Gitlin, *Media Unlimited*, 2001, 27. Howard Rheingold adds, "We're in a period where the cutting edge of change has moved from the technology to the literacies made possible by the technology." Rheingold, *Net Smart*, 2012, 3. Lisa Gitelman adds, "Amateur newspapers, fanzines, and their successors have always been imagined in contrast to commercially published periodicals." Gitelman, *Paper Knowledge*, 2014, 149.

13. Quoted in Ensminger, "The Allure of the Instant," 2012, 18.

14. Interview with the author, February 6, 2023.

15. Quoted in Munzenrider, "Starting a New Skateboard Magazine and Other Radical Acts of Love," 2023.

16. Interview with the author, April 30, 2024.

17. Skateboarding is one of the best examples of the oft-quoted line from Automatic Jack in William Gibson's 1982 short story "Burning Chrome": "The street finds its own use for things." W. Gibson, "Burning Chrome," 1982, 106. See also Liu, *The Laws of Cool*, 2004, 308.

18. Borden, *Skateboarding, Space and the City*, 2001.

19. Rosenberger, *Callous Objects*, 2017, 11.

20. Davis, *City of Quartz*, 1990, 233.

21. Norman, *Turn Signals Are the Facial Expressions of Automobiles*, 1992, 43.

22. For example, see "Skateboard Deterrents," Park Warehouse, https://parkwarehouse.com/product-category/other/skateboard-deterrents.

23. N. Smith and Walters, "Desire Lines and Defensive Architecture," 2018, 2980–95.

24. Andreou, "Anti-Homeless Spikes," 2015. Toby Heys calls Muzak "the sound of the working dead." Heys, "Muzak and the Working Dead," in Goodman, Heys, and Ikoniadou, *AUDINT*, 2019, 283. Jonathan Crary writes, "Public spaces are now comprehensively planned to deter sleeping, often including—with an intrinsic cruelty—the serrated design of benches and other elevated surfaces that prevent a human body from reclining on them." Crary, *24/7*, 2013, 26–27.

25. See Beachy, *The Most Fun Thing*, 2021, 38; and Zarka, *On a Day with No Waves*, 2011, 71.

26. Interview by the author, November 2, 2022. Being a skateboarder is as much about who you aren't as it is who you are. See Cliver, *Shit*, 2016, 144.

27. See Beachy, *The Most Fun Thing*, 38–40; and "Drunks at Osiris," 2007.

28. Johan Kugelberg compares skateboarders to the Indigenous Australians in Bruce Chatwin's *Songlines*. See Kugelberg, foreword to *I'll Kick You in the Head with My Energy Legs*, by Jonnie Craig, 2012.

29. Interview with the author, November 2, 2022.

30. Louison, *The Impossible*, 2011; Zarka, *On a Day with No Waves*, 2011.

31. In Hill, *The Man Who Souled the World*, 2007.

32. Quoted in Brooke, *Concrete Wave*, 1999, 43.

33. Louison, *The Impossible*, 2011, 155. World Industries' disruption was another "cultural spasm that [could] only exhaust itself," as Simon Reynolds writes of no wave, *Rip It Up and Start Again* (157).

34. Quoted in Jonze, "Steve Rocco Interview," 1992, 26–27.

35. Quoted in Christopher, "Tod Swank: Foundation's Edge," in *Follow for Now*, 2007, 270.

36. Ibid.

37. Ibid., 274.

38. Ibid., 272.

39. "Advertising is a perfect—yes, essential—medium for a meta-production that is no longer geared towards a production of goods but instead works by selling services and buying activities. It is not simply a plateau of persuasion where we are goaded into buying stuff. Instead, it is a passage to the real." Andreasen and Larsen, *The Critical Mass of Mediation*, 2014, 75.

40. Boyle, *Authenticity*, 2003, 12.

41. Rushkoff, *Life, Inc.*, 2009, 119.

42. Interview with the author, November 2, 2022.

43. Boyle, *Authenticity*, 2003, 106.

44. Naisbitt, Naisbitt, and Philips, *High Tech/High Touch*, 1999, 31.

45. Interview with the author, November 2, 2022.

46. Beachy, *The Most Fun Thing*, 2021, 13.

47. Wiles, *Plume*, 2019, 174.

48. "How 130-Year-Old National Geographic Became a Top Social Media Brand," 2018.

49. Zarka, *On a Day with No Waves*, 2011, 37.

50. Mike Vallely cites this as one of the reasons *Thrasher* was more authentic in its representation of skateboarding. "*Thrasher* was skateboarding to me," he says. Quoted in Mortimer, *Stalefish*, 2008, 117. *Transworld* was published in print from 1983 through 2019 and is now a web-only publication. See O'Haver, "An Oral History of 'Transworld Skateboarding' Magazine," 2019.

51. Zarka, *On a Day with No Waves*, 2011, 41; O'Haver, "An Oral History of 'Transworld Skateboarding' Magazine," 2019.

52. Quoted in O'Haver, "An Oral History of 'Transworld Skateboarding' Magazine," 2019. Zero owner and skateboard pro Jamie Thomas adds, "Then *Thrasher* had a limited distribution, but every single issue was consistently amazing." Quoted in ibid.

53. Louison, *The Impossible*, 2011, 47.

54. Interview with the author, November 8, 2019.

55. Ibid.

56. Ibid.

57. Interview with the author, November 11, 2019.

58. Quoted in Cohen, "Highlights from Ian MacKaye's Library of Congress Lecture," 2013.

59. O. Howell, "The Poetics of Security," 2001.

CHAPTER 6. A MESSAGE IN A BOTTLENECK

Epigraphs are drawn from the following sources: Jeter quote from *Noir: A Novel* (New York: Bantam Books, 1998), 375. Herbert quote from *Dune Messiah* (New York: Putnam, 1969), 232. Morton quote from *The Ecological Thought* (Cambridge, Mass.: Harvard University Press, 2010), 80. Calvino quote from *Six Memos for the Next Millennium* (Cambridge, Mass.: Harvard University Press, 1988), 43. Ortega y Gasset quote from *The Dehumanization of Art and Ideas about the Novel* (Princeton: Princeton University Press, 1925), 33. Dick quote from "How to Build a Universe That Doesn't Fall Apart Two Days Later," in *I Hope I Shall Arrive Soon* (New York: St. Martins, 1985), 9. Bridle quote from *New Dark Age* (New York: Verso, 2018), 13. Golding quote from *Lord of the Flies* (New York: Penguin, 1954), 27. Bradbury quote from *Bradbury Speaks* (New York: William Morrow, 2005), 153. Jung quote from *The Undiscovered Self* (Princeton, N.J.: Princeton University Press, 1957/1990), 24. VanderMeer quote from *Annihilation* (New York: Farrar, Strauss and Giroux, 2014), 66. Erickson quote from *Days Between Stations* (New York: Owl Books, 1985), 269. Nash, Thorsen, and Kunzelmann, from "Slogans," in *Cosmonauts of the Future: Texts from the Situationist Movement in Scandinavia and Elsewhere*, ed. Mikkel Bolt Rasmussen and Jakob Jakobsen (Copenhagen, Denmark: Nebula, 2015), 129. Postman quote from *Amusing Ourselves to Death* (New York: Penguin, 1985), 15.

1. Drawn by Jim Philips, this was the first such series on a skateboard deck. See B. Marcus, *The Skateboard*, 2011, 204.

2. Kimmerer, *Braiding Sweetgrass*, 2013, 46 (thanks to Michael Schandorf for sharing this with me). As the philosopher Philip Wheelwright put it, "The transmutive process that is involved may be described as *semantic motion*; the idea of which is implicit in the very word 'metaphor,' since the motion (*phora*) that the word connotes is a semantic motion—the double imaginative act of outreaching and combining that essentially marks the metaphoric process." Wheelwright, *Metaphor and Reality*, 1962, 71–72.

3. McLuhan and McLuhan, *Laws of Media: The New Science*, 1988, 3. Eric McLuhan notes, "Metaphor and the tetrad on metaphor are the very heart of *Laws of Media*." McLuhan and McLuhan, *The Lost Tetrads of Marshall McLuhan*, 2017, 200n. So much of Marshall McLuhan's work was done with metaphors. As he wrote, interpolating Robert Browning, "A man's reach must exceed his grasp or what's a metaphor." McLuhan, *Understanding Media*, 1964, 64. See also Van Den Eede, "Exceeding Our Grasp," in Rogers, Whalen and Taylor, *Finding McLuhan*, 2015, 43–61; and Logan, *McLuhan Misunderstood*, 2013, 39–40.

4. Marshall McLuhan credits Kenneth Boulding with the "break boundary" concept. McLuhan, *Understanding Media*, 1964, 41.

5. Quoted in Christopher, "Eugene Thacker," in *Follow for Now*, 2007, 15–16. William Empson wrote, "Metaphor, more or less far-fetched, more or less complicated, more or less take for granted (so as to be unconscious), is the normal mode of development of a language." Empson, *Seven Types of Ambiguity*, 1947, 2. As Jedediah Purdy puts it, "Some words come to be like ghosts, roaming around among us without flesh, vaguely recalling old dreams and tragedies that the living have forgotten." Purdy, *For Common Things*, 1999, 77–78.

6. These are referred to as *frozen, historical,* or *dead* metaphors. See Lakoff and Johnson, *Metaphors We Live By*, 1980.

7. Prinz, *Art Discourse/Discourse in Art*, 1991, 144. See also Cage, *Silence*, 1961, 12.

8. See Lakoff, "The Contemporary Theory of Metaphor," in Ortony, *Metaphor and Thought*, 1993, 202–51; Raunig, *A Thousand Machines*, 2010; and Wilden, *System and Structure*, 1972.

9. Gimpel, *The Medieval Machine*, 1976, 147–48.

10. Botkin, *Discordant Harmonies*, 1990, 114; Vroon, "Man-Machine Analogs and Theoretical Mainstreams in Psychology," in Baker et al., *Current Issues in Theoretical Psychology*, 1987, 393–414; and Cormac, *A Cognitive Theory of Metaphor*, 1985. As Graham Dunstan Martin puts it, the human mind has been compared to "whatever contemporary technology happens to be lying about at the time." Martin, *Shadows in the Cave*, 1990, 10. See also Gentner and Grudin, "The Evolution of Mental Metaphors in Psychology," 1985, 181–91. As the philosopher La Mettrie put it in 1748, justifying the use of such metaphors, "man is so complicated a machine that it is impossible to get a clear idea of the machine beforehand and hence impossible to define it." La Mettrie, *Man, a Machine*, 1912, 89.

11. Knapp, *Machine, Metaphor, and the Writer*, 1989, 28. John Shotter writes, "We cannot avoid using narratives, metaphors or theories, but what we can avoid is becoming entrapped within their confines by claiming any one of them to be the single correct narrative, metaphor or theory. They are instruments, not depictions." See Shotter, *Conversational Realities*, 1993, 132.

12. Giedion, *Mechanization Takes Command*, 2013, 512.

13. Huxley, *America and the Future*, 1970, 10.

14. Galbraith, *The New Industrial State*, 1967, 19.

15. For a thorough and grounded exploration of this idea, see Bratton, *The Stack*, 2015. See also Huxley, *America and the Future*, 1970.

16. Quoted in Eshun, "William Gibson: The Co-Evolution of Humans and Machines," in Christopher, *Follow for Now, Vol. 2*, 2021, 282.

17. Le Corbusier, *The City of To-morrow and Its Planning*, 1978, 179–90. As J .G. Ballard put it, "The high-rise was a huge machine designed to serve, not the collective body of tenants, but the individual resident in isolation." Ballard, *High-Rise*, 1975, 17. See also Theil, *People, Paths, and Purposes*, 1997.

18. Manuel DeLanda writes, "As skyscraper designers know well, radical changes in form may be needed once a critical height has been reached, such as the use of an interconnected iron or steel frame, which, beginning in the 1850s, liberated walls from their load-bearing duties transforming them into mere curtains. Other components playing a material role are those determining the *connectivity* of regions of a building. If locales are stations where the daily paths of individual persons converge, the regions that subdivide them must be connected to each other to allow for the circulation of human bodies and a variety of other material entities." DeLanda, *A New Philosophy of Society*, 2006, 96. See also Vance, *The Continuing City*, 1990, 24–25.

19. M. Bateson, *Our Own Metaphor*, 1972, 284. See also G. Bateson, "Our Own Metaphor: Nine Years After," in *A Sacred Unity*, 1991, 285.

20. This is another trend that John Naisbitt spotted in newspapers in the late 1970s. See chapter 8, "From Hierarchies to Networks," in Naisbitt, *Megatrends*, 1982.

21. Galloway, Thacker, and Wark, *Excommunication*, 2014, 2. Extending it inward, Michael Schandorf writes that "every 'node' is a network all its own, each with its own very fuzzy boundaries and interpenetrations." Schandorf, *Communication as Gesture*, 2019, 108. Latour defines the black box similarly: "Each of the parts inside the black box is itself a black box full of parts." Latour, *Pandora's Hope*, 1999, 185. Expanding it outward, Barry Brummett writes: "[Texts are] nodal: what one experiences here and now is a text, but it may well be a part of a larger text extending into time and space. Texts tend to grow nodes off themselves that develop into larger, more complex but related texts." Brummett, *A Rhetoric of Style*, 2008, 118.

22. Shaviro goes on to flip McLuhan's claim that electronic networks are an extension of the central nervous system to write that "every individual brain is a miniaturized replica of the global communications network." Shaviro, *Connected*, 2003, 12. Albert Laslo-Barabasi says, "This universality of network structure and evolution tells us that if we understand one network, we can apply this knowledge to understand all other complex webs out there." Quoted in Christopher, "Albert Laslo-Barabasi: Think Networks," in *Follow for Now*, 2007, 105.

23. Nicholas Carr writes, "Those who celebrate the 'outsourcing' of memory to the Web have been misled by a metaphor. They overlook the fundamentally organic nature of biological memory. What gives real memory its richness and its character, not to mention its mystery and fragility, is its contingency. It exists in time, changing as the body changes. Indeed, the very act of recalling a memory appears to restart the entire process of consolidation, including the generation of proteins to form new synaptic terminals." Carr, *The Shallows*, 2010, 191.

24. John Brockman writes, "Not only are the metaphors we use to invent the universe always changing and transforming, but like biological entities, they are no doubt following their own still-undiscovered laws of evolution, their own survival of the fittest, and perhaps even their own versions of molecular drive and genetic drift." Brockman, *Einstein, Gertrude Stein, Wittgenstein*, 1986, 133. Brian Rotman adds, "Metaphors, narratives, and other interpretive linguistic modes we use for human sense making of the world around us do the work of conditioning us to behave as if we and the world were digital. Language and ideological productions thus serve as kinds of virus vectors preparing the ground for the gradual shift in ontology." Rottman, *Becoming Beside Ourselves*, 2008, xi.

25. Metcalfe's Law, named after the Xerox engineer Bob Metcalfe by George Gilder, states that the power of a network grows as the square of its number of users grows. That is, the cost of the network is linear while its value is exponential. Gilder, "Telecosm: Metcalfe's Law and Legacy," 1993.

26. Kevin Kelly calls this "the fax effect." Kelly, *New Rules for the New Economy*, 1998, 39–49.

27. Using metaphors from epidemiology, Malcolm Gladwell calls this fatigue "immunity." Gladwell, *The Tipping Point*, 2000, 271–75.

28. McLuhan and McLuhan, *Laws of Media*, 1988.

29. See ibid.; and McLuhan and McLuhan, *The Lost Tetrads of Marshall McLuhan*, 2017.

30. McLuhan and McLuhan, *Laws of Media*, 1988, 145.

31. Stephenson, *In the Beginning Was the Command Line*, 1999.

32. Gitelman, *Paper Knowledge*, 2014, 17.

33. Douglas Rushkoff writes, "Analog media such as radio and television were continuous, like the sound on a vinyl record. Digital media, by contrast, are made up of many discrete sam-

ples. Likewise, digital networks break up our messages into tiny packets and reassemble them on the other end. Computer programs all boil down to a series of 1s and 0s, on or off. This logic trickles up to the platforms and apps we use." Rushkoff, *Team Human*, 2019, 63.

34. "Ways they have been internalized": Gitelman, *Paper Knowledge*, 2014, 17.

35. Raley, "TXTual Practice," in Hayles and Pressman, *Comparative Textual Media*, 2013, 20.

36. Galloway, *The Interface Effect*, 2012.

37. Ibid., 25.

38. Ibid.

39. Galloway, *The Interface Effect*, 2012. "Encoded experiences": I stole this term from James Bridle. See Bridle, "Books Are Encoded Experiences," in Lovink, Gerritzen and Kampman, *I Read Where I Am*, 2011, 56.

40. Watts and Strogatz, "Collective Dynamics of 'Small World' Networks," 1998, 393, 440–42.

41. Mitchell, *Complexity*, 2009, 236–39. See also Strogatz, *Sync*, 2003; and Rothenberg, *Bug Music*, 2013.

42. Hoffmann et al., "Duets Recorded in the Wild Reveal," 2019.

43. Ma, Ter Maat, and Gahr, "Neurotelemetry Reveals Putative Predictive Activity," 2020.

44. Iacoboni, *Mirroring People*, 2008; Rizzolatti and Sinigaglia, *Mirrors in the Brain*, 2008; Byrne and Levitan, "The Seed Salon," 2007, 46.

45. Iacoboni, *Mirroring People*, 2008; Rizzolatti and Sinigaglia, *Mirrors in the Brain*, 2008. Dedre Gentner and Jonathan Grudin point out, however, "The causal connection between physiological organization [of the brain] and the organization of the mind, though irrelevant from the point of view of the metaphor, may lead to a decrease in vigilance." Gentner and Grudin, "The Evolution of Mental Metaphors in Psychology," 1985, 191.

46. Heim, *The Metaphysics of Virtual Reality*, 1993, 98.

47. Gibson, "Johnny Mnemonic," 1981, 98. Dominic Pettman asks, "What could be more human that wanting to control one's own trace?" Pettman, *Look at the Bunny*, 2011, 95. Bruno Latour laments, "It is as if the inner workings of private worlds have been pried open because their inputs and outputs have become thoroughly traceable." Latour, "Beware, Your Imagination Leaves Digital Traces," 2007. Jonathan Crary adds: "One effect of this imposition of an input/ouput model is a homogenization of inner experience and the contents of communication networks, and an unproblematic reduction of the infinite amorphousness of mental life to digital formats." Crary, *24/7*, 2013, 98.

48. Bush, "As We May Think," 1945, 101–8. Ted Nelson writes that "the written word is nothing less than the tracks left by the mind." See Nelson, *Computer Lib/Dream Machines*, 1987, 50.

49. Norman, *Living with Complexity*, 2011, 134–35.

50. Hagle, "Are You Smarter Than an Algorithm?," 2023. In Cormac McCarthy's *Stella Maris*, Alicia Western says, "I'm not sure that it's all that wise to commit things to memory. What you log in becomes fixed in a way that the machinations of the unconscious would appear not to." McCarthy, *Stella Maris*, 2022, 100. In K. W. Jeter's *Noir*, "the ultimate barfly" says, "In this world, memory's dead weight. We can do without it." Jeter, *Noir*, 1998, 331.

51. Hagle, "Are You Smarter Than an Algorithm?," 2023.

52. Quoted in Eshun, "William Gibson," in Christopher, *Follow for Now, Vol. 2*, 2021, 278.

John Brockman writes, "Search for rhythms and patterns, . . . The analysis moves from the study of fixed entities that are capable of ownership to the transaction of the species with environmental forces. Look to the transaction." Brockman, *Afterwords*, 1973, 9. See also Pettman, *Infinite Distraction*, 2016, 97.

53. W. Gibson, *Idoru*, 1996, 50.

54. Merholz, "Metadata for the Masses," 2004.

55. This method of tagging content to make it easier to find is called "ethnoclassification." See Star, "Slouching towards Infrastructure," 1996; and Merholz, "Ethnoclassification and Vernacular Vocabularies," 2004.

56. Walker, *Buying In*, 2009, 68.

57. De Certeau, *The Practice of Everyday Life*, 1984. As Russell Potter writes, "it hollows out a fallout shelter where the ostensible, 'official' significance of words and pictures is made shiftable, mutable, unreliable." Potter, *Spectacular Vernaculars*, 1995, 108. Erkki Huhtamo and Jussi Parikka add, "Media are in this sense a reservoir for tactics and techniques for manipulating humans and their culture." Huhtamo and Parikka, *Media Archaeology*, 2011, 25.

58. Bey, *T.A.Z.*, 1985, 5.

59. Becker, *Tactical Reality Dictionary*, 2004, 12. Adam Greenfield calls it a "discourse of seamlessness" that "deprives the user of meaningful participation in the decisions that affect his or her experience." Greenfield, *Everyware*, 2006, 137–38.

60. Branwyn, *Jamming the Media*, 1997, 248.

61. Dery, *Culture Jamming*, 1993. The semiotician Umberto Eco called culture jamming "semiological guerrilla warfare" in 1967. See Eco, *Travels in Hyperreality*, 1986, 135–44. Culture jammers are the hackers of what Richard Doyle calls "rhetorical software." See Hayles, *How We Became Posthuman*, 1999, 118.

62. Greil Marcus, *Lipstick Traces*, 1989, 179. Mary Douglas writes, "In a culture of mistrust, language itself becomes suspect and deceiving." Douglas, *Thinking in Circles*, 2007, 145.

63. Riley, *After the Mass-Age*, 2017, 58. See also Parser, *The Filter Bubble*, 2011; and Rushkoff, *Survival of the Richest*, 2022, 44–45. The practice has given way to what Clyde Wayne Crews once called "splinternets." See Kumar, "Libertarian, or Just Bizarro?," 2001.

64. Dominic Pettman writes that it's more accurately described as *attention ecologies*, plural, "each with its own ecosystem and microclimate." Pettman, *Infinite Distraction*, 2016, 95. See also Bourdieu, *On Television*, 1998.

65. Shaviro, *Connected*, 2003, 71.

66. Ibid.

67. M. Bateson, "How to Be a Systems Thinker," 2018.

68. Swigart, "A Writer's Desktop," in Laurel, *The Art of Human-Computer Interface Design*, 1990, 140–41.

CHAPTER 7. DISGUISE THE LIMIT

Part 3 epigraphs are drawn from the following sources: Deleuze and Guattari quote from *A Thousand Plateaus* (Minneapolis: University of Minnesota Press, 1987), 24–25. Bachelard quote from *The Poetics of Space* (Boston: Beacon Press, 1964), 12. Donnie Dark quote from the movie *Donnie Darko*, directed and written by Richard E. Kelly (Los Angeles: Newmarket Films, 2001).

Chapter 7 epigraphs are drawn from the following sources: Strode quote from *My Little Book of Prayer* (Chicago: Open Court, 1905). Toms quote from *Pacemaker* (Ballinlough, Cork: Banshee Press, 2022), 1. Williams quote from "Elohim (1972)," on *Black Whole Styles* [CD] by various artists (New York: Big Dada, 1998). Sukenick quoted in Jane Alison, *Meander, Spiral, Explode* (New York: Catapult, 2019), 6. Raunig quote from *Dividuum* (South Pasadena, Calif.: Semiotext(e), 2016), 71. McKay quoted in Andrea Kurland, "Getting Deep with Ian MacKaye, the Godfather of DIY Culture," *Huck Magazine* 49 (March/April 2015), May 27, 2020, https://www.huckmag.com/art-and-culture/music-2/ian-mackaye-survival-issue-interview/. Robinson quote from *2312: A Novel* (New York: Orbit, 2012). Jackson quote from *A Sense of Place, A Sense of Time* (New Haven, Conn.: Yale University Press, 1996), 190. Bakker quote in *Neuropath* (New York: Tor, 2009), 279. Lockwood quote in *No One is Talking About This* (New York: Riverhead, 2021, 143). France quoted in Bruce Chatwin, *The Songlines* (New York: Viking, 1987), 174. Gibson quote from *Count Zero* (London: Voyager, 1986), 165.

1. Shaviro, *Doom Patrols*, 1996, 114. See also Parikka, *Insect Media*, 2010.

2. See Paul, *Dinosaurs of the Air*, 2002; and Chiappe, *Glorified Dinosaurs*, 2007.

3. David Wells postulates, "There is technology as soon as there are limbs, as soon as there is bending of those limbs, as soon as there is any articulation at all. As soon as there is articulation, the human has rounded the technological bend, the technological turn has occurred, and there is no more simple human." Wells, *Dorsality*, 2008, 3.

4. Shipman, *Taking Wing*, 1998, 89.

5. See Schaller, "Wing Evolution," in Hecht et al., *The Beginnings of Birds*, 1985, 333–48.

6. See Chiappe, *Glorified Dinosaurs*, 2007. In butterflies, wings are used for thermoregulation in addition to flight. See also Halpern, *Four Wings and a Prayer*, 2001.

7. McLuhan wrote, "The transformations of technology gave the character of organic evolution because all technologies are extensions of our physical being." McLuhan, *Understanding Media*, 1964, 199.

8. For more on the history of early bicycles, see the introduction to D. Wilson, *Bicycling Science*, 2004, 1–35. McLuhan added, "Unlike the wing or fin, the wheel is lineal and requires road for its completion." McLuhan, *Understanding Media*, 1964, 198.

9. As John Lienhard puts it, "Do you remember your first bike or the first car you drove? Think back for a moment. That bike was like a magic carpet. It changed you, irrevocably." Lienhard, *The Engines of Our Ingenuity*, 2000, 19. Manuel DeLanda writes, "When a young child learns to swim or to ride a bicycle, . . . a new world suddenly opens up for experience, filled with new impressions and ideas. The new skill is deterritorializing to the extent that it allows the child to break with past routine by venturing away from home in a new vehicle, or inhabiting previously forbidden spaces like the ocean." DeLanda, *A New Philosophy of Society*, 2006, 50. Friedrich Kittler wrote, "The innocence that comes into being where bodies and media technologies come into contact is called flight of ideas." Kittler, *Discourse*, 1990, 238.

10. McLuhan, *Understanding Media*, 1964, 198–99. See also McLuhan and McLuhan, *The Lost Tetrads of Marshall McLuhan*, 2017, 102, which describes the tetrad of the bicycle: "Enhances locomotion. Obsolesces walking. Retrieves equilibrium. Flips into airplane."

11. For more details on the Wright brothers' story, their bicycles, and flying machines, see

Gibbs-Smith, *The Aeroplane*, 1960; J. Anderson, *Inventing Flight*, 2004; and Perry, *Bike Cult*, 1995, 143–47.

12. McMahon and Graham, *Introduction to Engineering Materials*, 1992, 4. Peter L. Jakab adds, "The Wrights are commonly described as bicycle mechanics turned airplane builders, yet the highly influential role bicycles played in their inventive work is rarely emphasized." Jakab, *Visions of a Flying Machine*, 1990, 7. James Means wrote in 1896, "Wheeling is just like flying. To learn to wheel one must learn to balance; to learn to fly one must learn to balance." Means, "Wheeling and Flying," 1896, 24–25.

13. Wolff, "Flight," 1978.

14. For a discussion of this connection, see Kyl and Gronen, "The Bicycle-Airplane Connection," 1990, 88.

15. John Durham Peters writes, "The body is the most basic of all media, and the richest with meaning, but its meanings are not principally those of language or signs." Peters, *The Marvelous Clouds*, 2015, 6.

16. Eno, "The Revenge of the Intuitive," 1999, 176.

17. See Sterling, "Bicycle Repairman," in Kelly and Kessell, *Rewired: The Post-Cyberpunk Anthology*, 2007, 15.

18. As Michael Heim points out, "There are far more clues to the underlying structural processes for the person riding a bicycle than there are for a person writing on a computer screen." Heim, *Electric Language*, 1993, 131–32.

19. Seymour, "Foreword," in Long, *Walking the Line*, 2002, 9.

20. Wagner, *Symbols That Stand for Themselves*, 1986, 21.

21. Ingold, *Lines*, 2007, 79, emphasis in original. See also McCloud, *Understanding Comics*, 1993, esp. 118–37.

22. Solnit, *Wanderlust*, 2001, 5. See also Kagge, *Walking*, 2019; and Gros, *A Philosophy of Walking*, 2014.

23. De Certeau, *The Practice of Everyday Life*, 1984, 97–98.

24. Ibid., 99.

25. J. Michael Thomson wrote, "Walking is essential to the urban environment." Thomson, *Great Cities and Their Traffic*, 1977, 52.

26. De Certeau, *The Practice of Everyday Life*, 1984, 100.

27. McDonough, "Situationist Space," in *Guy Debord and the Situationist International*, 2002, 241–65. See also Prescott-Steed, "Frostbite on My Feet," 2013, 45–68.

28. Ingold, *Lines*, 2007, 15; O'Rourke, *Walking and Mapping*, 2013.

29. W. Gibson, *All Tomorrow's Parties*, 1999, 158–59.

30. Calvino, *Invisible Cities*, 1974, 11.

31. De Certeau writes, "The paths that correspond in this intertwining, unrecognized poems in which each body is an element signed by many others, elude legibility. It is as though the practices organizing a bustling city were characterized by their blindness. The networks of these moving, intersecting writings compose a manifold story that has neither author nor spectator, shaped out of fragments of trajectories and alterations of spaces." De Certeau, *The Practice of Everyday Life*, 1984, 93. See also Lynch, *The Image of the City*, 1960, 49; and Rapoport, *The Meaning of the Built Environment*, 1982, 115.

32. Desire lines are also known less poetically as "desire paths" and "social trails." See Lidwell, Holden, and Butler, *Universal Principles of Design*, 2010.

33. The use finds its own street for things. See Brand, *How Buildings Learn*, 1994, 187; Lidwell, Holden, and Butler, *Universal Principles of Design*, 2010, 76–77; Norman, *Living with Complexity*, 2010, 126–31; and Merholz, "Metadata for the Masses," 2004. Other examples include Australian engineers spreading flour and using the resultant immediately visible trails to improve path planning, as well as what Aaron Naparstek, a visiting scholar at MIT's department of urban studies and planning, calls "sneckdowns": the edges of paths left in snow. See P. H., "Undriven Snow," 2014; and Mars and Kohlstedt, *The 99% Invisible City*, 2020.

34. Norman, *Living with Complexity*, 2010, 128.

35. Smith and Walters, "Desire Lines and Defensive Architecture," 2018. See also Rogers, *Rebuilding Central Park*, 1987, 35.

36. Ingold, *Lines*, 2007, 43.

37. Lynch, *What Time Is This Place?*, 1972, 186.

38. "Well-worn ribbons of earth": Shepard and Murray, "Introduction: Space, Memory and Identity," in Murray, Shepard, and Hall, *Desire Lines*, 2007, 1. "Yearning": La Plante quoted in "Whose Sidewalk Is It, Anyway?" by Brown, 2003. La Plante, chief traffic engineer for engineering firm T. Y. Lin International, adds, "When sidewalks are provided, people do walk." Brown also touts the importance of the social encounters sidewalks enable, quoting Jane Jacobs's classic *The Death and Life of Great American Cities*, "Lowly, unpurposeful and random as they appear, sidewalk contacts are the small change from which a city's wealth of public life may grow." Brown, "Whose Sidewalk Is It Anyway?," 72. Brandon LaBelle notes, "The sidewalk is a threshold between an interior and an exterior, between different sets of rhythms that come to orchestrate the dynamic passing of exchange each individual body instigates and remains susceptible to." LaBelle, *Acoustic Territories*, 2010, 88.

39. Brighenti, "New Media and the Prolongations," 2010, 475. See also O'Rourke, *Walking and Mapping*, 2013.

40. See Byrne, *Bicycle Diaries*, 2009.

41. "Historical layers": Lynch, *What Time Is This Place?*, 1972, 186.

42. Ewen, *All Consuming Images*, 1988, 114.

43. See Scott, *Domination and the Arts of Resistance*, 1990.

44. Alexander et al., *The Oregon Experiment*, 1975, 65–66.

45. Theil, *People, Paths, and Purposes*, 1997. Donald Norman calls desire lines "social signifiers, a clear indication that people's desires do not match the vision of the planners." Norman, *Living with Complexity*, 2010, 126. See also Rapoport, *The Meaning of the Built Environment*, 1982.

46. Lynch and Hack, *Site Planning*, 1984, 205–6.

47. See Greil Marcus, *Lipstick Traces*, 1989; Wark, *50 Years of Recuperation of the Situationist International*, 2008.

48. Debord, "Theory of the Dérive," in Knabb, *Situationist International Anthology*, 1981, 50.

49. See Corner, "The Agency of Mapping," in Cosgrove, *Mappings*, 1999, 213–52; and Wark, *The Beach Beneath the Street*, 2011.

50. The Situationists saw the dérive as a way to make the city playful. See Careri, *Walkscapes*, 2002, esp. 68–118. If you want to see the result of taking Debord and the *dérive* too seriously, see Nicholson, *The Lost Art of Walking*, 2008, 142–71.

51. Greil Marcus, "Heading for the Hills," 1999.

52. See Waxman, *Keep Walking Intently*, 2017, esp. 83–190.

53. Careri, *Walkscapes*, 2002, 102. He is describing Debord's *The Naked City: Illustration de l'hypothèse des plaques tournantes en psychogéographique* (Copenhagen: Permild and Rosengreen, n.d. [May 1957]), lithographed sheet in black and red, 47 x 33 cm.

54. "In search of signs": Greil Marcus, *Lipstick Traces*, 1989, 168.

55. Sommer, *Personal Space*, 1969, 41. See also Norman, *Living with Complexity*, 2010.

56. Foucault, *The History of Sexuality*, vol. 1, 1980.

57. Hampton and Cole, *Soft Paths*, 1988, 26–27.

58. Solnit, *Wanderlust*, 2001, 213.

59. Brighenti, "New Media and the Prolongations of Urban Environments," 2010, 475. One can also view bottom-up as tactical and the top-down as strategic. See de Certeau, *The Practice of Everyday Life*, 1984.

60. Solnit, *Wanderlust*, 2001, 213.

61. See the collection of images of desire lines from all over in the Desire Path Flickr Group, http://www.flickr.com/groups/desire_paths.

62. Tilley, *A Phenomenology of Landscape*, 1997, 30.

63. Hampton and Cole, 1988, 27.

64. Lynch, *Soft Paths*, 1960, 3. Kevin Lynch's idea of the "legible city" is so widely accepted as to have become a part of the vocabulary of urban studies. See Scott, *Seeing Like a State*, 1998. James Holston writes, "Thus, the semantic structure of the street organizes the entire cityscape into a coherent and predictable order." Holston, *The Modernist City*, 1989, 123. E. V. Walter refers to the city as a "symbolic pattern." Walter, *Placeways*, 1988, 5.

65. See Norman, *Living with Complexity*, 2010, 131–36. Tilley writes,

> First, points linked by a clear path have achieved a degree of structural homology and hence positive cultural identity. The points linked by paths share sets of common elements—sacred stones, trees, artefact depositions, names and titles referred to in myths and stories and linked to the activities of ancestors who stopped on the journey which created the path. Second, linked places on paths can be understood in terms of sequential precedence, a hierarchy of ancestral origin points from which paths radiate to others. Priority in time is linked to the ceremonial precedence and power of places linked by paths. Third, paths structure experiences of the places they link, they help to establish a sense of linear order. A path brings forth possibilities for repeated actions within prescribed confines.

Tilley, *A Phenomenology of Landscape*, 1997, 30, paraphrasing Parmentier, *The Sacred Remains*, 1987, 109–11.

66. See Carl Myhill's widely cited but reckless application of the term "desire line" to everything from poor mapping of stove controls to ergonomic keyboard design. Myhill, "Commercial Success by Looking for Desire Lines."

67. Lynch, *The Image of the City*, 1960, 3.

68. Alexander, *Notes on the Synthesis of Form*, 1964, 88. John Zacharias writes, "It has long been observed that people prefer wider pathways in the absence of other meanings that might attach to choices. Spaciousness may be a fundamental characteristic desired by pedestrians, although it is also likely that certain meanings attach to spatial dimensions, including the importance of the pathway in the network." Zacharias, "The Pedestrian Itinerary," in Timmermans, *Pedestrian Behavior*, 2009, 298. See also Herzog and Flynn-Smith, "Preference and Perceived Danger," 2001. Jane Alison opens her beautiful book *Meander, Spiral, Explode* with an anecdote about a furniture designer named Eileen Gray, who in 1926 built a house based on the paths most followed by her and her housekeeper. "She made diagrams of their motions and those of the sun to reveal natural patterns," Alison writes, "loops in the kitchen, deep lines by the windows, meanders through the living room." She built a house based on this "organic choreography." Alison, *Meander, Spiral, Explode*, 2019, 3.

69. Mattern, *Code+Clay . . . Data+Dirt*, 2017.

70. Rose, *Interstate*, 1990, 15. See also Daniels and Warnes, *Movement in Cities*, 1980.

71. In 1896 Sir Robert Hunter cited UK Highway Act, 1835, 5 & 6 Will. IV, cap. 50, Sections 84–92: "The only ground on which a public footpath can be diverted is that the proposed footpath is more commodious for the public than the existing footway." Hunter, *The Preservation of Open Spaces*, 1896, 393.

72. Throgmorton and Eckstein, "Desire Lines."

73. See A. Black, "The Chicago Area Transportation Study," 1990; Creighton, *Urban Transportation Planning*, 1980.

74. In a previous plan devised around the turn of the century, one public official stated his appreciation that they had pursued policies "for Chicago's development and security, in the present and future, along lines of beauty and convenience." See Smith, *The Plan of Chicago*, 2006, 68.

75. Scott, *Seeing Like a State*, 1998, 56.

76. Hales, "Grid, Regulation, Desire Line," in Orville and Meikle, *Public Space and the Ideology of Place in American Culture*, 2009, 167.

77. See Kelling and Coles, *Fixing Broken Windows*, 1996; and Wilson and Kelling, "Broken Windows," 1982.

78. See Sampson and Raudenbush, "Systematic Social Observation of Public Spaces," 1999.

79. Scott, *Seeing Like a State*, 1998, 53.

80. Ibid., 73; Ingold, *Lines*, 2007, 116.

81. Turchi, *Maps of the Imagination*, 2004, 28. See also O'Connor, *Wayfinding*, 2019, esp. 98–127. Bruce Chatwin wrote, "Our early explorations are the raw materials of our intelligence." Chatwin, *Anatomy of Restlessness*, 1996, 100.

82. Long, *Walking the Line*, 2002, 68.

83. Roelstraete, *Richard Long*, 2010, 2. Rudi H. Fuchs calls it "a fundamental interruption of the history of art." See Fuchs, *Richard Long*, 1986, 43.

84. Quoted in Craig-Martin, "The Artist Michael Craig-Martin Talks with Richard Long," 2009.

85. In Long, *Heaven and Earth*, 2009, 172.

86. Long, *Selected Statements and Interviews*, 2007, 25. Tim Ingold writes, "Footprints are individual; paths are social." Ingold, *The Life of Lines*, 2015, 63.

87. Ingold, *The Life of Lines*, 2015, 66.

88. McLuhan wrote, "The road is our major architectural form." McLuhan, *Culture Is Our Business*, 1970, 132. Laurie Anderson says, "A Road is always a good metaphor for a plot." Quoted in King, "Laurie Anderson in Conversation," 2018, 14.

89. O'Rourke, *Walking and Mapping*, 2013, xvii.

90. Long, *Selected Statements and Interviews*, 2007, 25.

91. Long, *Walking the Line*, 2002; Nicholson, *The Lost Art of Walking*, 2008.

92. Long, *Selected Statements and Interviews*, 2007, 73.

93. Michell, introduction to Watkins, *The Old Straight Track*, 1970, xv–xvi.

94. Michell, *The New View over Atlantis*, 1995, 101. See also Morrison, *Pathways to the Gods*, 1978.

95. Michell, *The New View over Atlantis*, 1995, 104. See also Cowan and Silk, *Ley Lines of the UK and USA*, 2013, esp. 117–22.

96. Lynch, *What Time Is This Place?*, 1972, 1.

97. Ibid.

CHAPTER 8. LOCATION IS EVERYWHERE

Epigraphs are drawn from the following sources: Apologies to Mark Wieman, from whom I stole this title. Mooney quotes from *Easy Travel to Other Planets: A Novel* (New York: Farrar Straus & Giroux, 1981), 219, 59. McLuhan quote from *Culture Is Our Business* (New York: Ballantine Books, 1970), 12. Superstudio quoted in John Beckmann, *The Virtual Dimension* (New York: Princeton Architectural Press, 1998). VanderMeer quotes from *Acceptance* (New York: Farrar, Straus and Giroux, 2014), 79, 81. El Akkad quote from *American War* (New York: Alfred A. Knopf, 2017), 164.

1. Calvino, *Invisible Cities*, 1974, 35. Paul Virilio added, "Thanks to satellites, the cathode-ray window brings to each viewer the light of another day and the presence of the antipodal place. If space is that which keeps everything from occupying the same place, this abrupt confinement brings absolutely everything precisely to that 'place,' that location that has no location. The exhaustion of physical, or natural, relief and of temporal distances telescopes all localization and all position. As with live televised events, the places become interchangeable at will." Virilio, *The Lost Dimension*, 1991, 17–18.

2. Garreau, *Edge City*, 1991. As William Cronen wrote, "Cities were the stars around which town and country satellites would come to orbit." See Cronen, *Nature's Metropolis*, 1991, 39.

3. Norton, "Feral Cities," 2003.

4. Brand, *How Buildings Learn*, 1994, 178. Pierre Bourdieu calls designers "cultural intermediaries." See Bourdieu, *Distinction*, 1984.

5. Quoted in Smith, *Conversations with William Gibson*, 2014, 222. As social animals our first need is to communicate. See Douglas and Ney, *Missing Persons*, 1998, esp. 46–73.

6. See Chatwin, *Anatomy of Restlessness*, 1996, esp. 85–99; Krims, *Music and Urban Geography*, 2007, 34.

7. Byrne, *Bicycle Diaries*, 2009.

8. Borchert writes, "The settlement machine is distinctive . . . because its internal parts include people, whose innovations make the machine self-redesigning in evolutionary, unpre-

dictable ways." Borchert, "Futures of American Cities," in Hart, *Our Changing Cities*, 1991, 218 and throughout.

9. Daniels and Warnes, *Movement in Cities*, 1980.

10. Cronen refers to waterways as "natural advantages" for cities. See Cronen, *Nature's Metropolis*, 1991, 55. Steven Johnson writes, "For millennia, most cities had been bound inexorably to the natural ecosystem that lay outside their walls: the energy flowing through the fields and forests around them established a population ceiling they couldn't grow beyond." Johnson, *The Ghost Map*, 2006, 93.

11. "Frictionless space": Vance, "Human Mobility and the Shaping of Cities," in Hart, *Our Changing Cities*, 1991, 75. John R. Borchert writes, "American places are nodes at junctions of routes in the circulation network." Borchert, "Futures of American Cities," 1991, 219.

12. Bruce Chatwin wrote, "The City is a sheepfold superimposed over a Garden." Chatwin, *The Songlines*, 1987, 203.

13. Borchert, "Futures of American Cities," 1991, 238.

14. Mailander, *LA at Intermission,* 2014.

15. Quoted in Montgomery, *Happy City*, 2013, 7.

16. Ibid., 9.

17. Ibid., 75.

18. Quoted in Goodyear, "An Illustrated History of All," 2014.

19. Ibid.

20. Earnest, "The Poetic Politics of Space," 2014.

21. Yanow, *War of Streets and Houses*, 2014, 23.

22. Ibid., 20, 21.

23. Harvey, *Rebel Cities*, 2012, 161–62.

24. Mailander, *LA at Intermission*, 2014, 75.

25. Chatwin, *Anatomy of Restlessness*, 1996, 89.

26. McLuhan, *Understanding Media*, 1964, 374–75. Alan Liu writes,

> In an age when communications media increasingly supplanted physical transportation in carrying the bulk of exchanges between firms and their public, suppliers, contractors, and regulatory agencies, business necessarily had to be conducted in a compound idiom of information and service we might call *informed service*. That is, information—its timely collection, storage, processing, management, and delivery, whether conducted by a firm itself or outsourced to specialized accounting, legal, public relations, courier, and other service firms—did not simply facilitate service: it now *was* the ultimate service. For a firm to make effective contact across its boundary with other sectors or organizations, armies of salesmen were no longer adequate or, in many cases, even possible. Instead, firms had to work the telephone, the TV, the newspaper, and so on, which meant cumulatively that they had to transmit an ever larger proportion of both their image and products as information.

Liu, *The Laws of Cool*, 2004, 119.

27. McLuhan and McLuhan, *Laws of Media*, 1988, 202; Florida, *The Great Reset*, 2010.

28. Greenfield, *Against the Smart City*, 2013. The French philosopher Gilles Deleuze

called these "any-space-whatever," the space in his view only important for the connections it facilitates. Deleuze, *Cinema 1*, 1983. See also Jacobs, *The Economy of Cities*, 1969, esp. chap. 4. In a letter to György Kepes, Marshall McLuhan wrote, "I think we are already living in a new kind of world city. . . . Since we are actually living in that new city electrically, it is inevitable that the perceptual life of the young will be accommodated to it in spite of the irrelevance of the concepts that they are being taught." McLuhan, *Letters of Marshall McLuhan*, 1987, 453. Stewart Brand's "city planet" is already here; it is just not evenly distributed yet.

29. Brighenti, "New Media and the Prolongations," 2010.

30. Virilio, *The Art of the Motor*, 1995, 140. As Douglas Rushkoff puts it, "Our technologies change from being the tools humans use into the environments in which humans function." Rushkoff, *Team Human*, 2019, 52. See also Berry and Dieter, *Postdigital Aesthetics*, 2000.

31. Sassen, *The Global City*, 1991, 5, 3–4. See also Friedmann, "The World City Hypothesis," 1986.

32. Sassen, *The Global City*, 1991, 10.

33. See Wallace-Wells, "William Gibson," 2011; W. Gibson, *Spook Country*, 2007, 63–64.

34. James E. Vance Jr. writes, "In the absence of any previous development, current urban morphologies will reflect the dominant current transportation." Vance, "Human Mobility and the Shaping of Cities," 1991, 83.

35. I stole this term from artist Peter Root. See Caroline Williamson, "Peter Root's *Ephemicropolis*—A City of Staples," *Design Milk*, August 8, 2012, https://design-milk.com /peter-roots-ephemicropolis-a-city-of-staples.

36. Interview with the author, February 6, 2002.

37. See Kilcullen, *Out of the Mountains*, 2013.

38. Lynch, *What Time Is This Place?*, 1972, 21.

39. Kluitenberg, "The Society of the Unspectacular," 2007. See also Kluitenberg, *Delusive Spaces*, 2008.

40. Walter, *Placeways*, 1988.

41. Lynch, *Managing the Sense of a Region*, 1976.

42. Pettman, *In Divisible Cities*, 2013, 3.

43. "Inhabitants" seemed too restrictive a term for the tactical media activity envisioned here, so I opted for "participants."

44. De Certeau, *The Practice of Everyday Life*, 1984; Brighenti, "New Media and the Prolongations of Urban Environments," 2010, 482.

45. Murphy, *Last Futures*, 2016, 69.

46. McLuhan, *Understanding Media*, 1964, 114. He also wrote, "By putting our physical bodies inside our extended nervous systems, we set up a dynamic by which all previous technologies that are mere extensions of hands and feet and teeth and bodily heat-controls—all such extensions of our bodies, including cities—will be translated into information systems." McLuhan, *Understanding Media*, 1964, 63.

47. Ford and Vaucher, "Lockdown and Breakout," in Butt, Eshun, and Fisher, *Post-Punk*, 2016, 129. See also Ford, *Savage Messiah*, 2011.

48. Ford says, "I've been moving away from psychogeography into what I've started to call a 'sociogeography'. So it's not just about your own memories and experiences; it's about tuning

into collective moments of intensity, the psychic tremors or the buried currents and narratives that exist in places." In Ashman, "A Q&A with Laura Oldfield Ford," 2017. Benjamin Bratton writes, "Cities are media for the circulation of potentials (as well as the encapsulation of foregone conclusions) and to search that potential this means getting out of our own skins." See Bratton, *The New Normal*, 2017, loc. 142, Park Books.

49. Kodwo Eshun writes, "The suburb incubates a certain kind of boredom that functions as a precondition for dreaming your way into the future. In a contemporary era characterized by nested interruptions, that kind of boredom has to be artificially created." In Butt, Eshun, and Fisher, *Post-Punk*, 2016, 21.

50. McLuhan, *Understanding Media*, 1964, 221. "The city is obsolete," Marshall McLuhan contended, "Ask the computer. In the computer age, speech yields to gesticulation or the direct interface of total cultures." McLuhan, *Counterblast*, 1969, 12–13. Nicholas Negroponte writes, "Digital living will include less and less dependence upon being in a specific place at specific time, and the transmission of place itself will start to become possible." Negroponte, *Being Digital*, 1995, 163. He also writes, "In the post-information age, since you may live and work at one or many locations, the concept of an 'address' now takes on new meaning." Ibid., 238. See also Mattern, "Sidewalks of Concrete and Code."

51. Peters, *The Marvelous Clouds*, 2015, 6, 266–70. Mary Douglas writes, "The body, as a vehicle of communication, is misunderstood if it is treated as a signal box, a static framework emitting and receiving strictly coded messages. The body communicates information for and from the social system in which it is a part." Douglas, *Implicit Meanings*, 1975, 83. Walter Ong wrote, "The body is a frontier between myself and everything else." Ong, *Orality and Literacy*, 1982, 73. Judith Donath writes, "We are embodied beings, who have evolved in the physical world; our thoughts and imagination are rooted in the sensory experience of our physical surroundings. Online, there is no body; there is only information. We comprehend abstract ideas by reframing them in metaphoric terms that ultimately derive from physical experience." Donath, *The Social Machine*, 2014, 9;. In Alan Liu's *The Laws of Cool*, he describes William Gibson's metaphor for the body: "What gives such surface effect substance is the deep feeling Gibson instills in *superficies* for the generalized destructivity of society—for the way dominant social, economic, political, and military institutions use technology to hollow out people and habitats to leave empty shells of identity." Liu's writing is specifically about Gibson's *Neuromancer* here. Liu, *The Laws of Cool*, 2004, 336. See also ibid., 339; Douglas, *Natural Symbols*, 1970, esp. 65–81; and Waite, *Mediation and the Communication Matrix*, 2003, 76–77.

52. McLuhan wrote, "Media effects are new environments as imperceptible as water to a fish, subliminal for the most part." McLuhan, *Counterblast*, 1969, 22.

53. Viola, *Reasons for Knocking at an Empty House*, 1995.

54. Ito, "The Future Form of Visual Art," 1994, 69. See also Milutis, *Ether*, 2005. Channeling dolphin spirits, June Hughes writes, "We would have you contemplate the possibility or notion, that the sleek and fluid bodies that you see in motion through the waves—these dynamic forms of seeming perfection in your realm are satellites, or what have become known amongst many of you as space ships, star ships, manifested in your third dimensional reality and piloted by star beings that some term extra-terrestrial, and that we prefer to call angelic. And they zip and glide through your oceans, these ships of living light as poetry in motion,

as rhythmic waves, light particles condensed in form." Hughes, *Whale Wisdom, Dolphin Joy*, 2014, 19.

55. The musician and musicologist Bernie Krause calls the collective sounds of plants and animals "biophony." See Giggs, *Fathoms*, 2020, 152.

56. See Grodin and Lindlof, *Constructing the Self in a Mediated World*, 1996; Jenkins, *Convergence Culture*, 2006; and Van Dijck, *The Culture of Connectivity*, 2013.

57. Papacharissi, *A Networked Self*, 2011, 305.

58. Ito, "The Future Form of Visual Art," 1994, 69. Ted Mooney writes, "Dolphins are students of the sonic, the tidal, and the gravitational. Through ear and skin, the dolphin receives forty million bits of such information per second and organizes them spontaneously into a changing musical replica of the world. Some of this music is useful; some is not." Mooney, *Easy Travel to Other Planets*, 1981, 59.

59. Postman, "The Reformed English Curriculum," in Eurich, *High School 1980*, 1970, 161. See also Peters, *The Marvelous Clouds*, 2015; and Mattern, *Code+Clay . . . Data+Dirt*, 2017. In his 1970 book *Expanded Cinema*, Gene Youngblood called ours the *Paleocybernetic Age*. Erik Davis writes, "For Youngblood, expanded cinema meant nothing more or less than expanded consciousness, the drive—spiritual as well as technological—to manifest the spectral machinery of mind in the world before our eyes." E. Davis, *Beyond Belief*, 2003, 21.

60. Chatwin, *Anatomy of Restlessness*, 1996, 88.

61. Ibid., 98.

62. The most famous example of this is described in Ray Bradbury's novel *Fahrenheit 451* (New York: Ballantine Books, 1953). The filmmaker Arthur Jafa tells Jace Clayton,

> Nam June Paik once said that the culture which will survive in the future is the one you can carry around in your head. Black culture is a profound demonstration of that, because we're strong, traditionally speaking, in the space of what I call immaterial invention, or immaterial expressivity. We came from Africa, where there are 1,000-year-old traditions of material expressivity—sculpture, architecture, etc. But, in the US, we've been defined by the Middle Passage. On a slave ship, all you can take with you is song or rhetoric. You can't take a sculpture or a building with you on a chain gang or to prison. So, that tradition of immaterial expressivity continues. A mix is a perfect example of this because it involves taking pre-existing material and creating new relationships through proximity.

Clayton, *Uproot*, 2016. See also Eshun, *More Brilliant Than the Sun*, 1998, 192.

CHAPTER 9. SUTURE SELF

Epigraphs are drawn from the following sources: Eagleton quote from *After Theory* (New York: Basic Books, 2003), 208. Cadigan quote from "The Final Remake of *The Return of Little Latin Larry* with a Completely Remastered Soundtrack and the Original Audience," in *Rewired: The Post-Cyberpunk Anthology*, ed. James Patrick Kelly and John Kessell (San Francisco, Calif.: Tachyon Publications, 2007), 98. Indian proverb quoted in Bruce Chatwin, *The Songlines* (New York: Viking, 1987), 181. Kluitenberg quote from *Delusive Spaces: Essays on Culture, Media and Technology* (Rotterdam, Netherlands: NAi, 2008), 283. Baudrillard quote from *Cool Memories, IV: 1995–2000* (London: Verso, 2003), 64. Gray quote from *Straw*

Dogs (New York: Farrar, Straus and Giroux, 2002), 196. Ishiguro quote from *Klara and the Sun* (New York: Knopf, 2021), 293.

1. McLuhan, *Understanding Media*, 1964; Kittler, *Optical Media*, 2010, 29.

2. Gordon, "Laurie Anderson," 1980, 51–54.

3. L. Anderson, *Stories from the Nerve Bible*, 1994, 48. Anderson was writing here about her piece "The Handphone Table." Anderson wired up an ordinary table that listeners could lean on with their elbows, place their hands over their ears, and hear sounds through their bones.

4. Quoted in Bahadur, "Laurie Anderson," 1984, 9.

5. See Hegarty, *Rumour and Radiation*, 2015, 96; and González, Gordon, and Higgs, *Christian Marclay*, 26. In 1976. Anderson also had a "viophonograph," a violin with a turn-table mounted on it to play 7" records with a bow equipped with a needle. See Block and Glasmeier, *Broken Music*, 85.

6. Official [UK] Singles Chart Top 75, October 18, 1981–October 24, 1981, https://www .officialcharts.com/charts/singles-chart/19811018/7501. When the budget for the National Endowment for the Arts was cut in the 1980s, Anderson's unlikely pop career was partially a product of her not being able to find funding. A&R representative Tim Carr said, "I started to bring bands up to record labels because you could get Laurie Anderson a record deal where you could no longer get her a grant." Quoted in LeRoy and Relic, *For Whom the Cowbell Tolls*, 42.

7. Dery, "Signposts on the Road to Nowhere," in DeCurtis, *Present Tense*, 1992, 149–66.

8. Dery, "Laurie Anderson," 1989, 76.

9. Anderson says, "Fortunately for me there's the word 'state' in the title which is handy, just in terms of your mental and emotional state and how things tie together to give some sort of *impression*." Quoted in Black, "Altered States," 1984, 18.

10. Quoted in Howell, *Laurie Anderson*, 1992, 10.

11. See Hegarty, *Rumour and Radiation*, 2015, 89–91. Douglas Kellner writes, "Her art is an erotics of surfaces and the play of light, sound, movement, word, and performance." Kellner, *Cultural Studies, Identity, and Politics*, 1995, 289.

12. Quoted in Duckworth, *Talking Music*, 1995, 382.

13. Ibid.

14. Quoted in J. Howell, *Laurie Anderson*, 1992, 101.

15. See Duckworth, *Talking Music*, 1995.

16. Quoted in King, "Laurie Anderson in Conversation," 2018, 34.

17. Quoted in McCorduck, "America's Multi-Mediatrix," 1994, 137.

18. See Cubitt, "Laurie Anderson," in Roberts, *Art Has No History*, 1994, 278–96; Reed, *Laurie Anderson's "Big Science,"* 2022; L. Anderson and Marclay, "All the Right Notes," 60–65.

19. McCorduck, "America's Multi-Mediatrix," 1994, 137.

20. Rapaport, "Can You Say Hello?," 353. See also Rapaport, *Between the Sign and the Gaze*, 268–92; and Braidotti, *Nomadic Subjects*, 259–61.

21. Benjamin, "The Work of Art in the Age of Its Technological Reproducibility," 39. As Mary Douglas and Aaron Wildavsky wrote, "It is good to realize that the concepts of center and border are entirely abstract and relative to the discussion. The sense of border is inherent in the consciousness of the people who perceive their lives as uncommitted and essentially

critical of some defined other part of human society where power resides." Douglas and Wildavsky, *Risk and Culture*, 1983, 103.

22. Youngblood, *Expanded Cinema*, 58; Seabrook, *Nobrow*, 2000.

23. Li and Lai, "Voice, Object and Listening," 347–61.

24. Quoted in King, "Laurie Anderson," 18.

25. L. Anderson, "Words in Reverse," 6.

26. Morton, *The Ecological Thought*, 103.

27. Braidotti, *Nomadic Subjects*, 279. See also L. Anderson, *Stories from the Nerve Bible*, 1994.

28. Kellner, *Cultural Studies*, 388.

29. Davis, "Recording Angels," in Young, *Undercurrents*, 2002, 17–18.

30. Parikka, *Insect Media*, 2010.

31. This is what Humberto Maturana and Francisco J. Varela call "structural coupling." Maturana and Varela, *The Tree of Life*, 1987. See also Maturana and Poerkson, *From Being to Doing*, 2004.

32. Parikka, *Insect Media*, 2010, xxvii. See also G. Bateson, *Steps to an Ecology of Mind*, 1972.

33. Quoted in Elterman, "Bret Easton Ellis Has All the Answers," 2022.

34. Nietzsche wrote, "The oversaturation of an age with history seems to me to be hostile and dangerous to life . . . it leads an age into a dangerous mood of irony in regard to itself and subsequently into the even more dangerous mood of cynicism: in this mood, however, it develops more and more a prudent practical egoism through which the forces of life are paralyzed and at last destroyed." Nietzsche, *Untimely Meditations*, 1997, 5.

35. Kaufmann, *The Comedian as Confidence Man*, 1997.

36. Wampole, "How to Live Without Irony," 2012.

37. Purdy, *For Common Things*, 1999.

38. As Purdy put it in 1999, "They are not so much ushering in the next millennium as riding out the last." Ibid., 95.

39. Greenwald, *Nothing Feels Good*, 2003, 55.

40. Quoted in Wampole, "How to Live Without Irony," 2012.

41. J. Butler, *Precarious Life*, 2004, 30.

42. Ibid.

43. Quoted in Sacks, *Poking a Dead Frog*, 2014, 38.

44. All of these tribulations may seem trivial, but, as Jaron Lanier writes, "pop culture is important." He adds, "It drags us all along with it; it is our shared fate. We can't simply remain aloof." Lanier, "Where Did the Music Go?," in Miller, *Sound Unbound*, 2008, 385.

45. Reynolds, *Retromania*, 2011, 419. See also Purdy, *For Common Things*, 1999.

46. Reynolds, *Retromania*, 2011, 419–20.

47. W. Gibson, "The Recombinant City," in Delaney, *Dhalgren*, 2001, xiii.

48. Wampole writes, "Moving away from the ironic involves saying what you mean, meaning what you say and considering seriousness and forthrightness as expressive possibilities, despite the inherent risks." Wampole, "How to Live Without Irony," 2012.

49. Braidotti, *Transpositions*, 2006, 153.

50. Rushkoff, *Present Shock*, 2013, 74. Roisin Kiberd writes, "It speaks to the brain's endless

search for novelty; the thing we want is always hidden, just off the screen, so we keep looking, dopamine spiking, brain held on the brink of receiving a reward." Kiberd, *The Disconnect*, 2021, 138. Paul Levinson characterizes text messaging on phones similarly. See Levinson, *Digital McLuhan*, 1999.

51. Quoted in Pescovitz, "Douglas Rushkoff," 2013.

52. Rushkoff, *Present Shock*, 2013, 2. James Gleick summed it up nicely when he told me in 1999, "We know we're surrounding ourselves with time-saving technologies and strategies, and we don't quite understand how it is that we feel so rushed. We worry that we gain speed and sacrifice depth and quality. We worry that our time horizons are foreshortened—our sense of the past, our sense of the future, our ability to plan, our ability to remember." Interview with the author, June 12, 1999.

53. Morrissey and Marr, "How Soon Is Now?," 1984.

54. Gergen, *The Saturated Self*, 6.

55. Michael Harris says, "Before all memory of those absences is shuttered, though, there is this brief time when we might record what came before. We might do something with those small, barely noticeable instances when we're reminded of our love for absence." Harris, *The End of Absence*, 2014, 8. Iain Chambers writes, "With electronic reproduction offering the spectacle of gestures, images, styles, and cultures in a perpetual collage of disintegration and reintegration, the 'new' disappears into a permanent present. And with the end of the 'new'—a concept connected to linearity, to the serial prospects of 'progress,' to 'modernism'—we move into a perpetual recycling of quotations, styles, and fashions: an uninterrupted montage of the 'now.'" Chambers, *Popular Culture*, 1986, 190. Michael Leyton says we are all "prisoners of the present." Leyton, *Symmetry, Causality, Mind*, 1992, 1.

56. Douglas Rushkoff writes, "Even though we may be able to be in only one place at a time, our digital selves are distributed across every device, platform, and network onto which we have cloned our virtual identities." See Rushkoff, *Present Shock*, 2013, 72. Brian Rotman writes, "This projected virtual user is a ghost effect: an abstract agency distinct from any particular embodied user, a variable capable of accommodating any particular user within the medium." Rotman, *Becoming Beside Ourselves*, 2008, xiii. And Judith Donath says, "The stranger, as we think of him now, may cease to exist." Donath, *The Social Machine*, 2014, 336.

57. "Doormen of discussions": Donath, *The Social Machine*, 2014, 159.

58. Kodwo Eshun writes, "The technological conditions for intervention in the present have to be artificially constructed. They are not spontaneously available. To embark on a project that is set in the present, you have to renounce digital abundance by undergoing a temporal diet or media famine. You have to turn yourself into a castaway marooned on an island of the present separated from the abundance of digital archives and previous musical eras that continually saturate the contemporary." In Butt, Eshun, and Fisher, *Post-Punk*, 2016, 20.

59. See Douglas, *How Institutions Think*, 1986, 92.

60. Ullman, *Close to the Machine*, 1997, 85.

61. Ibid. See also Baudrillard, *Seduction*. Sherry Turkle uses the term "seduction" in this context as well. Turkle, *Life on the Screen*, 1995.

62. Ullman, *Close to the Machine*, 1997, 89. Roisin Kiberd writes, "As if in response to the

grandiose nature of their never-ending mission, companies track the productivity of their workers with the same technologies they sell." Kiberd, *The Disconnect*, 2021.

63. Sherry Turkle writes, "The system presents us with what it believes we will buy or read or vote for. It places us in a particular world that constrains our sense of what is out there and what is possible." Turkle, *Reclaiming Conversation*, 2015, 307. Douglas Rushkoff says, "We must not accept any technology as the default solution for our problems. When we do, we end up trying to optimize ourselves for our machines, instead of optimizing our machines for us." Rushkoff, *Team Human*, 2019, 125.

64. Morville, *Ambient Findability*, 2005, xi.

65. David Toop writes, "One of the ways in which we negotiate our environment is through a refined awareness of resonance and its atmospheres." Toop, *Haunted Weather*, 2004, 63.

66. Morton, *The Ecological Thought*, 103.

67. McCullough, *Ambient Commons*, 2013, 20. Brian Eno describes ambient music as "constructing geography and not populating it." Eno foreword to Prendergast, *The Ambient Century*, 2000, vii. Eno also described it as "the colour of the light or the sound of the rain" and "as ignorable as it is interesting." Quoted in Prendergast, *The Ambient Century*, 2000, 115.

68. Quoted in Taraska, "Jewel Personality," 1995, 19.

69. See Kroker, *Spasm*, 1993; Sterling, "October 14, 1998, Beyond the Beyond," 2019.

70. See Wallace, *Both Flesh and Not*, 2007, 301.

71. Marshall McLuhan wrote, "The effect of electronic technology had at first been anxiety. Now it appears to create boredom." McLuhan, *Understanding Media*, 1964, 29. See also Harkaway, *The Blind Giant*, 2012, 49.

72. L. Anderson, "The Language of the Future," 1984. See also L. Anderson, *United States*, 1984, 24.

73. W. Gibson, *Distrust That Particular Flavor*, 2012, 51.

74. This is Régis Debray calls the "middle realm" between the technical and the social. Debray, *Media Manifestos*, 1996. Anderson adds, "In my work I strive for a sort of stereo effect, a pairing of things up against each other and see myself as a sort of a moderator between things." Quoted in Prinz, *Art Discourse/Discourse in Art*, 1991, 134.

75. L. Anderson, *Nothing in My Pockets*, 2009, 34.

76. Bey, *Immediatism*, 7. Bey continued, "The tendency of Hi Tech, & the tendency of Late Capitalism, both impel the arts farther and farther into extreme forms of mediation. Both widen the gulf between the production & consumption of art, with a corresponding increase in 'alienation.'" Ibid., 8.

77. Chinn and Roberts, "I, Ambient," in Rushkoff, *The GenX Reader*, 1994. Victor Turner describes the "betwixt and between" of ritual as "the peculiar unity of the liminal: that which is neither this nor that, and yet is both." Turner, *The Ritual Process*, 1969, 99.

78. Grant Morrison writes, "We have to leave our bodies and our cities behind us and go into space, just like the little fishes had to leave the sea was all they knew." Morrison, *The Invisibles*, 1994, 86.

79. Lund, "Laurie Anderson Interview," 2016. See also Dworkin, *Helicography*, 2021, 141–42.

80. As the Tactical Media Crew puts it, "Tactical Media are what happens when cheap 'do it yourself' media made possible by the revolution in consumer electronics, are exploited by those who are outside of the normal hierarchies of power and knowledge." See Tactical Media Crew, [2002?], https://tmcrew.org/enghome.htm. See also Raley, *Tactical Media*, 2009, 16.

81. Lanois, *Soul Mining*, 2010, 140.

82. Ibid. Jane Bennett writes, "Not only is human agency always already distributed in tools, microbes, minerals, and sounds, it only emerges as agentic by way of a distribution into the 'foreign' materialities its bearers are eager to exclude." Bennett, "The Agency of Assemblages," 2005, 463. Manuel DeLanda writes, "Human history is a narrative of contingencies, not necessities, of missed opportunities to follow different routes of development, not of a unilinear succession of ways to convert energy, matter, and information into cultural products." DeLanda, *A Thousand Years of Nonlinear History*, 1997, 99.

83. Morton, *The Ecological Thought*, 2010, 104.

84. Eno, foreword to Prendergast, *The Ambient Century*, 2000, xii.

85. As Gary Thompson puts it, "from the perspective of a raindrop there is no such thing as a cloud." Thompson, "Reweaving the Fabric of the Internet," 2011.

BIBLIOGRAPHY

Akkad, Omar El. *American War*. New York: Knopf, 2017.

Alexander, Christopher. *Notes on the Synthesis of Form*. Cambridge, Mass.: Harvard University Press, 1964.

Alexander, Christopher, Murray Silverstein, Shlomo Angel, Sara Ishikawa, and Denny Abrams. *The Oregon Experiment*. New York: Oxford University Press, 1975.

Alison, Jane. *Meander, Spiral, Explode: Design and Pattern in Narrative*. New York: Catapult, 2019.

Anderson, Chris. *The Long Tail: Why the Future of Business Is Selling Less of More*. New York: Hyperion, 2006.

Anderson, John D. *Inventing Flight: The Wright Brothers and Their Predecessors*. Baltimore: Johns Hopkins University Press, 2004.

Anderson, Laurie. "The Language of the Future," in *United States Live* [LP]. New York: Warner, 1984.

———. *Stories from the Nerve Bible: A Retrospective 1972–1992*. New York: Harper Perennial, 1994.

———. *Nothing in My Pockets*. Paris, France: Dis Voir, 2009.

———. *United States*. New York: Harper & Row, 1984.

———. "Words in Reverse." *Top Stories* (Buffalo, N.Y.), no. 2 (1979).

Anderson, Laurie, and Christian Marclay. "All the Right Notes." *ArtReview*, June 2005, 60–65.

Andreasen, Søren, and Lars Bang Larsen. *The Critical Mass of Mediation*. Copenhagen: Internationalistisk Ideale, 2014.

Andreou, Alex. "Anti-Homeless Spikes: 'Sleeping rough opened my eyes to the city's barbed cruelty.'" *Guardian*, February 18, 2015.

Antrim, Donald. *Elect Mr. Robinson for a Better World*. New York: Vintage, 1993.

Arthur, W. Brian. *The Nature of Technology: What It Is and How It Evolves*. New York: Simon & Schuster, 2009.

Ashman, Lydia. "A Q&A with Laura Oldfield Ford, Artist and Urban Explorer." *a-n*, February 15, 2017. https://www.a-n.co.uk/news/a-qa-with-laura-oldfield-ford-artist-and-urban-explorer.

Attali, Jacques. *Noise: The Political Economy of Music*. Minneapolis: University of Minnesota Press, 1977.

Augustine, Saint. *Confessions*. New York: Penguin, 1961.

Azerrad, Michael. *Our Band Could Be Your Life: Scenes from the American Indie Underground 1981–1991*. New York: Little, Brown, 2001.

Babbage, Charles. *Passages from the Life of a Philosopher*. New Brunswick: Rutgers University Press, 1994. First published 1864.

Bachelard, Gaston. *The Poetics of Space*. Translated from the French by Maria Jolas. Boston: Beacon, 1964.

Bagdikian, Ben H. *The Media Monopoly*. New York: Beacon, 1983.

Bahadur, Raj. "Laurie Anderson: Artist at Work." *Scene*, May 10–16, 1984.

Bakker, R. Scott. *Neuropath*. New York: Tor, 2009.

Ball, Jared. *I Mix What I Like! A Mixtape Manifesto*. Oakland: AK Press, 2011.

Ballard, J. G. *High-Rise*. London: Jonathan Cape, 1975.

Barfe, Louis. *Where Have All the Good Times Gone? The Rise and Fall of the Record Industry*. New York: Atlantic, 2004.

Barnes, Susan B. "Understanding Social Media from the Media Ecological Perspective." In *Mediated Interpersonal Communication*, edited by Elly A. Konijn, Sonja Utz, Martin Tanis, and Susan B. Barnes. New York: Routledge, 2008.

Barnett, Kyle. *Record Cultures: The Transformation of the U.S. Recording Industry*. Ann Arbor: University of Michigan Press, 2020.

Bartscherer, Thomas, and Roderick Coover, eds. *Switching Codes: Thinking Through Digital Technology in the Humanities and the Arts*. Chicago: University of Chicago Press, 2011.

Basile, Jonathan. *Tar for Mortar: The Library of Babel and the Dream of Totality*. Santa Barbara, Calif.: Punctum Books, 2018.

Bateson, Gregory. "Our Own Metaphor: Nine Years After." In *A Sacred Unity: Further Steps to an Ecology of Mind*. New York: HarperCollins, 1991.

———. *Steps to an Ecology of Mind*. New York: Ballantine, 1972.

Bateson, Mary Catherine. "How to Be a Systems Thinker: A Conversation with Mary Catherine Bateson." *Edge*, April 17, 2018. https://www.edge.org/conversation/mary_catherine_bateson-how-to-be-a-systems-thinker.

———. *Our Own Metaphor: A Personal Account of a Conference on the Effects of Conscious Purpose on Human Adaptation*. New York: Knopf, 1972.

Baudrillard, Jean. *Fragments: Cool Memories III, 1991–95*. London: Verso, 1997.

———. *Cool Memories, IV: 1995–2000*. London: Verso, 2003.

———. *Seduction*. London: Palgrave Macmillan, 1991.

———. *Simulacra and Simulations*. Stanford: Stanford University Press, 1988.

BBVA Foundation. "Interview with Mark Granovetter, Frontiers of Knowledge Award in Social Sciences." YouTube, April 7, 2022. https://www.youtube.com/watch?v=IMDFajzyYsI.

Beachy, Kyle. *The Most Fun Thing: Dispatches from a Skateboard Life*. New York: Grand Central, 2021.

Becker, Konrad. *Tactical Reality Dictionary: Cultural Intelligence and Social Control*. Brooklyn: Autonomedia, 2004.

Beckmann, John. *The Virtual Dimension: Architecture, Representation, and Crash Culture*. New York: Princeton Architectural Press, 1998.

Bennett, Jane. "The Agency of Assemblages and the North American Blackout." *Public Culture* 17, no. 3 (2005): 445–66.

Benjamin, Walter. *The Arcades Project*. Cambridge, Mass.: Belknap Press of Harvard University Press, 1999.

———. "The Work of Art in the Age of Mechanical Reproduction." In *Illuminations*. London: Fontana, 1968.

Berden, Frank Viander. "Robert Smith: 'I'm getting old, I feel empty, and I'm having so much fun.'" Translation of interview with Robert Smith from *HUMO* (Belgium), April 1989, 214–17. *The Cure: Pictures of You.* http://www.picturesofyou.us/89 /89-4-HUMO%20Belgium.htm.

Berry, David M., and Michael Dieter. *Postdigital Aesthetics: Art, Computation, and Design*. Houndmills, Basingstoke, UK: Palgrave-MacMillan, 2000.

Bey, Hakim. *Black Fez Manifesto, &c*. Brooklyn: Autonomedia, 2008.

———. *Immediatism*. San Francisco: AK Press, 1994.

———. *T.A.Z.: The Temporary Autonomous Zone, Ontological Anarchy, Poetic Terrorism*. Brooklyn: Autonomedia, 1985.

Birkenstein, Jeff, Anna Froula, and Karen Randell, eds. *The Cinema of Terry Gilliam: It's a Mad World*. New York: Columbia University Press, 2013.

Bitner, Jason. *Cassette from My Ex: Stories and Soundtracks of Lost Loves*. New York: St. Martin's, 2009.

Black, Alan. "The Chicago Area Transportation Study: A Case for Rational Planning." *Journal of Planning Education and Research* 10 (1990): 27–37.

Black, Bill. "Altered States." *Sounds* (London), December 22, 1984.

———. "Laurie Anderson: A Voice from Outer Space." *Sounds* (London), June 7, 1986.

Block, Ursula, and Michael Glasmeier, eds. *Broken Music: Artists' Recordworks*. Berlin: Berliner Kunstlerprogramm des DAAD, 1989.

Bloom, Allan D. *The Closing of the American Mind*. New York: Simon & Schuster, 1987.

Bloom, Howard. *The Lucifer Principle: A Scientific Expedition into the Forces of History*. New York: Atlantic Monthly, 1995.

Bofop, C. S. J. "*Thursday Afternoon*." Liner notes to Brian Eno's *Thursday Afternoon* (E.G. Records, 1985). Hyperreal Music Archive. http://music.hyperreal.org/artists/brian_eno /TA-txt.html.

Booker, M. Keith. *Comics Through Time: A History of Icons, Idols, and Ideas*, vol. 3, *1980–1995*. Santa Barbara, Calif.: Greenwood, 2014.

Borchert, John R. "Futures of American Cities." In *Our Changing Cities*, edited by John Fraser Hart. Baltimore: Johns Hopkins University Press, 1991.

Borden, Iain. *Skateboarding, Space and the City: Architecture and the Body*. Oxford, UK: Berg, 2001.

Botkin, Daniel B. *Discordant Harmonies: A New Ecology for the Twenty-First Century*. New York: Oxford University Press, 1990.

Bourdieu, Pierre. *Distinction: A Social Critique of the Judgement of Taste*. Cambridge, Mass.: Harvard University Press, 1984.

———. *The Logic of Practice*. Stanford: Stanford University Press, 1990.

———. *On Television*. New York: New Press, 1998.

Bown, Alfie. *Dream Lovers: The Gamification of Relationships*. London: Pluto, 2022.

Boyd, Katrina G. "Pastiche and Postmodernism in *Brazil*." *Cinefocus* 1, no. 1 (1990): 33–42.

Boyle, David. *Authenticity: Brands, Fakes, Spin and the Lust for Real Life*. New York: Harper Perennial, 2003.

Boym, Svetlana. *The Future of Nostalgia*. New York: Basic, 2001.

Bradbury, Ray. *Bradbury Speaks: Too Soon from the Cave, Too Far from the Stars*. New York: William Morrow, 2005.

———. *Fahrenheit 451*. New York: Ballantine, 1953.

Braidotti, Rosi. *Nomadic Subjects: Embodiment and Sexual Difference in Contemporary Feminist Theory*. New York: Columbia University Press, 1994.

———. *Transpositions: On Nomadic Ethics*. Cambridge, UK: Polity, 2006.

Brand, Stewart. *The Clock of the Long Now*. New York: Basic, 1999.

———. *How Buildings Learn: What Happens After They're Built*. New York: Viking, 1994.

Branwyn, Gareth. *Jamming the Media: A Citizen's Guide to Reclaiming the Tools of Communication*. San Francisco: Chronicle Books, 1997.

Bratton, Benjamin H. *The Stack: On Software and Sovereignty*. Cambridge, Mass.: MIT Press, 2015.

Bratton, Benjamin H, Nicolay Boyadjiev, and Nick Axel, eds. *The New Normal*. Moscow, Russia: Strelka, 2017.

Bridle, James. "Books Are Encoded Experiences." In *I Read Where I Am: Exploring New Information Cultures*, edited by Geert Lovink, Mieke Gerritzen, and Minke Kampman. Amsterdam: Valiz/Graphic Design Museum, 2011.

———. *New Dark Age: Technology and the End of the Future*. New York: Verso, 2018.

Brighenti, Andrea Mubi. "New Media and the Prolongations of Urban Environments." *Convergence* 16, no. 4 (2010), 471–87.

Brockman, John. *Afterwords: Explorations of the Mystical Limits of Contemporary Reality*. New York: Anchor, 1973.

———. *Einstein, Gertrude Stein, Wittgenstein: Re-Inventing the Universe*. New York: Viking, 1986.

Bromberg, Craig. *The Wicked Ways of Malcolm McLaren*. New York: HarperCollins, 1989.

Brooke, Michael. *Concrete Wave*. Toronto: Warwick, 1999.

Brooks, Dan. "In the 90s, We Worried About Nirvana 'Selling Out'. I Wish That Concept Still Made Sense." *Guardian*, July 18, 2023.

Brown, Patricia Leigh. "Whose Sidewalk Is It, Anyway?" *New York Times*, January 5, 2003.

Brummett, Barry. *Rhetoric of Machine Aesthetics*. Westport: Praeger, 1999.

———. *A Rhetoric of Style*. Carbondale: Southern Illinois University Press, 2008.

Buchloh, Benjamin H. D. "Andy Warhol's One-Dimensional Art: 1956–1966." In *Andy Warhol*, edited by Annette Michelson. Cambridge, Mass.:MIT Press, 2001

Bull, Michael. "The Audio-Visual iPod." In *The Sound Studies Reader*, edited by Jonathan Sterne. New York: Routledge, 2012.

———. "Automobility and the Power of Sound." *Theory, Culture, and Society*, 21, no. 4/5, 243–59.

———. "Investigating the Culture of Mobile Listening: From Walkman to iPod." In *Consuming Music Together: Social and Collaborative Aspects of Music Consumption Technologies*, edited by Kenton O'Hara and Barry Brown. Dordrecht, Netherlands: Springer, 2006.

———. *Sounding Out the City: Personal Stereos and the Management of Everyday Life*. Oxford: Berg, 2000.

———. *Sound Moves: iPod Culture and Urban Experience*. New York: Routledge, 2007.

Burgess, Anthony. *A Clockwork Orange*. London: William Heinemann, 1962.

Burningham, Bruce R. *Tilting Cervantes: Baroque Reflections on Postmodern Culture*. Nashville: Vanderbilt University Press, 2008.

Bush, Vannevar. "As We May Think." *Atlantic*, July 1945, 101–8.

Butler, Judith. *Precarious Life: The Powers of Mourning and Violence*. New York: Verso, 2004.

Butler, Samuel. *Erewhon, or Over the Range*. New York: Lancer, 1877.

Butt, Gavin, Kodwo Eshun, and Mark Fisher, eds. *Post-Punk: Then and Now*. London: Repeater, 2016.

Byrne, David. *Bicycle Diaries*. New York: Viking, 2009.

Byrne, David, and Daniel Levitan. "The Seed Salon." *Seed Magazine*, no. 10 (June 2007), 45–50.

Cadigan, Pat. "The Final Remake of *The Return of Little Latin Larry* with a Completely Remastered Soundtrack and the Original Audience." In *Rewired: The Post-Cyberpunk Anthology*, edited by James Patrick Kelly and John Kessell. San Francisco: Tachyon Publications, 2007.

———. "Rock On." In *Mirrorshades: The Cyberpunk Anthology*, edited by Bruce Sterling. New York: Arbor House, 1986.

Cage, John. *Silence: Lectures and Writings*. Middletown: Wesleyan University Press, 1961.

Calvino, Italo. *Invisible Cities*. New York: Harcourt, Brace, Jovanovich, 1974.

Campbell, Joseph *The Hero with a Thousand Faces*. New York: Pantheon, 1949.

Careri, Francesco. *Walkscapes: Walking as an Aesthetic Practice*. Barcelona: Editorial Gustavo Gili, 2002.

Carey, James W. *Communication as Culture: Essays on Media and Society*. New York: Psychology, 1992.

Carr, Nicholas. *The Shallows: What the Internet Is Doing to Our Brains*. New York: Norton, 2010.

Chambers, Iain. *Migrancy, Culture, Identity*. New York: Routledge, 1994.

———. *Popular Culture: The Metropolitan Experience*. New York: Routledge, 1986.

Chanan, Michael. *Repeated Takes: A Short History of Recording and Its Effects on Music.* London: Verso, 1995.

Chase, Barrett. "Philosophy of the Mix." *The Product*, October 14, 2005. Accessed November 5, 2024, https://web.archive.org/web/20060505165819/http://www.barrettchase.com/2005/10/philosophy_of_the_mix.html.

Chatwin, Bruce. *Anatomy of Restlessness: Uncollected Writings.* London: Jonathan Cape, 1996.

———. *The Songlines.* New York: Viking, 1987.

Chaw, Walter. *A Walter Hill Film.* Dallas: MZA Press, 2023.

Chiappe, Luis M. *Glorified Dinosaurs: The Origin and Early Evolution of Birds.* New York: Wiley, 2007.

Chinn, Meredith, and Todd C. Roberts. "I, Ambient." In *The GenX Reader*, edited by Douglas Rushkoff. New York: Random House, 1994.

Chow, Rey. "Listening Otherwise, Music Miniaturized: A Different Type of Question About Revolution." *Discourse* 13, no. 1 (1990): 129–48.

Christian, William. *The Idea File of Harold Adams Innis.* Toronto: University of Toronto Press, 1980.

"Christian Marclay" (interview). *Speed Kills* 3, no. 5 (1993), 21–29.

Christie, Ian. *Gilliam on Gilliam.* London: Faber & Faber, 1999.

Christopher, Roy, ed. *Follow for Now: Interviews with Friends and Heroes.* Seattle: Well-Red Bear, 2007.

———. *Follow for Now, Vol. 2: More Interviews with Friends and Heroes.* Santa Barbara, Calif.: Punctum Books, 2021.

Clayton, Jace. *Uproot: Travels in Twenty-First-Century Music and Digital Culture.* New York: FSG Originals, 2016.

Cliver, Sean, ed. *Shit: The Big Brother Book.* Berkeley: Gingko, 2016.

Cohen, Matt. "Highlights from Ian MacKaye's Library of Congress Lecture." *Spin*, May 8, 2013. https://www.spin.com/2013/05/fugazi-ian-mackaye-library-of-congress-lecture-punk-archive.

Constantino, Sara, and Stephen Nuñez. "Networks, Weak Ties, and Thresholds." *Phenomenal World*, August 8, 2019. https://www.phenomenalworld.org/interviews/networks-weak-ties-and-thresholds.

Corner, James. "The Agency of Mapping: Speculation, Critique and Invention." In *Mappings*, edited by Denis Cosgrove. London: Reaktion Books, 1999.

Corson, Ben. "Speed and Technicity (a Derridean Exploration)." PhD diss., Johns Hopkins University, 2000.

Coscarelli, Joe. "15 Songs in 15 Minutes: Inside Tierra Whack's Whimsical World." *New York Times*, June 6, 2018.

Coupland, Douglas. "Douglas Coupland on Generation X at 30: 'Generational trashing is eternal.'" *Guardian*, June 19, 2021.

———. *Generation X: Tales for an Accelerated Culture.* New York: St. Martin's, 1991.

———. *Marshall McLuhan: You Know Nothing of My Work!* New York: Norton, 2010.

Cowan, David R., and Chris Arnold. *Ley Lines and Earth Energies*. Kempton, Ill.: Adventures Unlimited, 2003.

Cowan, David R., and Anne C. Silk. *Ley Lines of the UK and USA*. Kempton, Ill.: Adventures Unlimited, 2013.

Craig-Martin, Michael. "The Artist Michael Craig-Martin Talks with Richard Long." *Independent*, May 8, 2009. http://peterfoolen.blogspot.com/2009/05/richard-long-michael -craig-martin.html.

Crary, Jonathan. *24/7: Late Capitalism and the Ends of Sleep*. New York: Verso, 2013.

Creighton, Roger L. *Urban Transportation Planning*. Urbana: University of Illinois Press, 1970.

Criqui, Jean-Pierre, ed. *On and by Christian Marclay*. Cambridge, Mass.: MIT Press, 2014.

Critchley, Simon. *ABC of Impossibility*. Minneapolis: Univocal, 2015.

Cronen, William. *Nature's Metropolis: Chicago and the Great West*. New York: Norton, 1991.

Cubitt, Sean. "Laurie Anderson: Myth, Management and Platitude." In *Art Has No History: The Making and Unmaking of Modern Art*, edited by John Roberts. New York: Verso, 1994.

Culkin, John M. "A Schoolman's Guide to Marshall McLuhan." *Saturday Review*, March 1967, 51–53, 70–72. Retrieved from *Unz Review*, http://www.unz.org/Pub/SaturdayRev -1967mar18-00051.

Cunningham, Andrew. "Sales of Vinyl Albums Overtake CDs for the First Time Since the Late '80s." *Ars Technica*, March 10, 2023. https://arstechnica.com/gadgets/2023/03/sales -of-vinyl-albums-overtake-cds-for-the-first-time-since-the-late-80s.

Daniels, P. W., and A. M. Warnes. *Movement in Cities: Spatial Perspectives on Urban Transport and Travel*. London: Methuen, 1980.

Danto, Arthur C. *Mark Tansey: Visions and Revisions*. New York: Abrams, 1992.

Davidson, Mark. *Uncommon Sense: The Life and Thought of Ludwig von Bertalanffy, Father of General Systems Theory*. Los Angeles: Tarcher, 1983.

Davis, Erik. *Beyond Belief: The Cults of Burning Man*. San Francisco: BRC, 2003.

———. "Recording Angels: The Esoteric Origins of the Phonograph." In *Undercurrents: The Hidden Wiring of Modern Music*, edited by Rob Young. London: Continuum, 2002.

Davis, Mike. *City of Quartz: Excavating the Future in Los Angeles*. New York: Verso, 1990.

Dawdy, Shannon Lee. *Patina: A Profane Archaeology*. Chicago: University of Chicago Press, 2016.

Dawson, Max. "Television Abridged: Ephemeral Texts, Monumental Seriality and TV-Digital Media Convergence." In *Ephemeral Media: Transitory Screen Culture from Television to YouTube*, edited by Paul Grainge. London: Palgrave MacMillan, 2011.

Debord, Guy. "Theory of the Dérive." In *Situationist International Anthology*, edited by Ken Knabb. Berkeley: Bureau of Public Secrets, 1981.

Debray, Régis. *Media Manifestos: On the Technological Transmission of Cultural Forms*. London: Verso, 1996.

De Certeau, Michel. *The Practice of Everyday Life*. Translated by Steven Rendall. Berkeley: University of California Press, 1984.

DeLanda, Manuel. *A New Philosophy of Society*. New York: Continuum, 2006.

———. *A Thousand Years of Nonlinear History*. Brooklyn: Zone, 1997.

Delany, Samuel R. *Dhalgren*. New York: Bantam, 1975.

Deleuze, Gilles. *Cinema 1: The Movement-Image*. Minneapolis: University of Minnesota Press, 1983.

Deleuze, Gilles, and Félix Guattari. *A Thousand Plateaus*. Minneapolis: University of Minnesota Press, 1987.

Denning, Michael. *Noise Uprising: The Audiopolitics of a World Musical Revolution*. London: Verso, 2015.

Derrida, Jacques. *Archive Fever: A Freudian Impression*. Chicago: University of Chicago Press, 1995.

Dery, Mark, *Culture Jamming: Hacking, Slashing, and Sniping in the Empire of Signs*. Open Pamphlet Series 23. Westfield, N.J.: Open Media, 1993.

———. *Escape Velocity: Cyberculture at the End of the Century*. New York: Grove, 1996.

———. "J. G. Ballard's Wild Ride." In *Extreme Metaphors: Collected Interviews*, edited by Simon Sellars and Dan O'Hara. London: Fourth Estate, 2012.

———. "Laurie Anderson." *Keyboard* 15, no. 12 (December 1989), 74–85, 88–90.

———. "The Mechanical Bridegroom Stripped Bare: A Catechism of McLuhanism for Nonbelievers." In *The Legacy of McLuhan*, edited by Lance Strate and Edward Wachtel. Cresskill, N. J.: Hampton Press, 2005.

———. "Signposts on the Road to Nowhere: Laurie Anderson's Crisis of Meaning." In *Present Tense: Rock and Roll and Culture*, edited by Anthony DeCurtis. Durham: Duke University Press, 1992.

DeSantis, Nick. "Radiohead's Digital Album Sales, Visualized." *Forbes*, May 10, 2016. http://www.forbes.com/sites/nickdesantis/2016/05/10/radioheads-digital-album-sales-visualized.

Dettmar, Kevin J. H. *33 1/3: Entertainment!* New York: Bloomsbury, 2014.

Deverson, Jane, and Charles Hamblett. *Generation X*. London: Tandem, 1964.

De Zengotita, Thomas. *Mediated: How the Media Shapes Our World and the Way We Live in It*. New York: Bloomsbury, 2005.

Dick, Philip K. "How to Build a Universe That Doesn't Fall Apart Two Days Later." In *I Hope I Shall Arrive Soon*. New York: St. Martins, 1985.

Donath, Judith. *The Social Machine: Designs for Living Online*. Cambridge, Mass.: MIT Press, 2014.

Douglas, Mary. *How Institutions Think*. Syracuse: Syracuse University Press, 1986.

———. *Implicit Meanings: Essays in Anthropology*. London: Routledge & Paul, 1975.

———. *Natural Symbols: Explorations in Cosmology*. London: Barrie & Rockliff, 1970.

———. *Purity and Danger: An Analysis of the Concepts of Pollution and Taboo*. London: Routledge & Paul, 1966.

———. *Thinking in Circles: An Essay on Ring Composition*. New Haven: Yale University Press, 2007.

Douglas, Mary, and Steven Ney. *Missing Persons: A Critique of Personhood in the Social Sciences*. Berkeley: University of California Press, 1998.

Douglas, Mary, and Aaron Wildavsky. *Risk and Culture: An Essay on the Selection of Technological and Environmental Dangers*. Berkeley: University of California Press, 1983.

"Drunks at Osiris: Osiris Teams Up with Consolidated Skateboards for the Drunk." *Transworld Skateboarding*, September 28, 2007. https://www.skateboarding.com/news /drunks-at-osiris.

Duckworth, William. *Talking Music: Conversations with John Cage, Philip Glass, Laurie Anderson, and 5 Generations of American Experimental Composers*. New York: Schirmer, 1995.

Du Gay, Paul, Stuart Hall, Linda Janes, and Keith Negus. *Doing Cultural Studies: The Story of the Sony Walkman*. London: Sage, 1997.

Dworkin, Craig. *Helicography*. Santa Barbara, Calif.: Punctum Books, 2021.

———. *No Medium*. Cambridge, Mass.: MIT Press, 2013.

Eagleton, Terry. *After Theory*. New York: Basic, 2003.

Earnest, Jarrett. "The Poetic Politics of Space: Rebecca Solnit in Conversation with Jarrett Earnest." *Brooklyn Rail*, March 4, 2014. https://brooklynrail.org/2014/03/art/rebecca -solnit-with-jarrett-earnest.

Easlea, Daryl. *Without Frontiers: The Life and Music of Peter Gabriel*. London: Omnibus, 2014.

Easterling, Keller. *Medium Design*. N.p.: Strelka, 2018.

Eco, Umberto. *Travels in Hyperreality: Essays*. New York: Harcourt Brace Jovanovich, 1986.

Egan, Jennifer. *A Visit from the Goon Squad*. New York: Knopf, 2010

Eichhorn, Kate. *Adjusted Margins: Xerography, Art, and Activism in the Late Twentieth Century*. Cambridge, Mass.: MIT Press, 2016.

Eisenstein, Sergei. "Montage of Attractions: For *Enough Stupidity in Every Wiseman*." *TDR: The Drama Review* 18, no. 1 (March 1974), 77–85.

Ellis, Bret Easton. *White*. New York: Knopf, 2019.

Elterman, Brad. "Bret Easton Ellis Has All the Answers." *Interview Magazine*, December 14, 2022. https://www.interviewmagazine.com/culture/bret-easton-ellis-has-all-the-answers.

Emerson, Lori. *Reading Writing Interfaces: From the Digital to the Bookbound*. Minneapolis: University of Minnesota Press, 2014.

Empson, William. *Seven Types of Ambiguity*. London: Chatto & Windus, 1947.

Eno, Brian. *A Year with Swollen Appendices*. London: Faber & Faber, 1996.

———. Foreword to Mark Prendergast, *The Ambient Century*, New York: Bloomsbury, 2000.

———. "The Revenge of the Intuitive: Turn Off the Options and Turn Up the Intimacy." *Wired*, January 1999, https://www.wired.com/1999/01/eno

Eno, Brian, and Russell Mills. *More Dark than Shark*. London: Faber & Faber, 1986.

Ensminger, David. "The Allure of the Instant: Postscripts from the Fading Age of Xerography." *Art in Print* 1, no. 6 (March–April 2012), 16–24.

Ernst, Wolfgang. *Digital Memory and the Archive*. Minneapolis: University of Minnesota Press, 2013.

Eshun, Kodwo. *More Brilliant than the Sun: Adventures in Sonic Fiction*. London: Quartet, 1998.

Eshun, Kodwo. "William Gibson: The Co-Evolution of Humans and Machines." In *Follow for Now, Vol. 2*, edited by Roy Christopher. Santa Barbara, CA: Punctum Books, 2021

Evens, Aden. *Sound Ideas: Music, Machines, and Experience*. Minneapolis: University of Minnesota Press, 2005.

Ewen, Stuart. *All Consuming Images*. New York: Basic, 1988.

Ferguson, Russell. "Never the Same Twice: In Conversation with Russell Ferguson, 2010." In *On and by Christian Marclay*, edited by Jean-Pierre Criqui. London: Whitechapel Gallery, 2014.

Fisher, Mark. *Capitalist Realism: Is There No Alternative?* London: Zero Books, 2009.

———. *The Ghosts of My Life: Writings on Depression, Hauntology, and Lost Futures*. London: Zero Books, 2014.

———. *K-Punk: The Collected and Unpublished Writings of Mark Fisher (2004–2016)*. London: Repeater, 2018.

Florida, Richard. *The Great Reset: How New Ways of Living and Working Drive Post-Crash Prosperity*. New York: Harper, 2010.

Fogarty, Mary. "Each One Teach One: B-Boying and Ageing." In *Ageing and Youth Cultures: Music, Style, and Identity*, edited by Andy Bennett and Paul Hodkinson. New York: Bloomsbury Academic, 2012.

Ford, Laura Oldfield. *Savage Messiah*. London: Verso, 2011.

Ford, Laura Oldfield, and Gee Vaucher, with Mark Fisher. "Lockdown and Breakout." In *Post-Punk*, edited by Gavin Butt, Kodwo Eshun, and Mark Fisher. London: Repeater Books, 2016.

Foster, Hal. "Slumming with Rappers at the Roxy." *London Review of Books* 22, no. 18 (September 21, 2000).

Foucault, Michel. *The History of Sexuality*, vol. 1, *An Introduction*. New York: Vintage, 1980.

Freud, Sigmund. "Fragment of an Analysis of a Case of Hysteria." In *The Standard Edition of the Complete Psychological Works of Sigmund Freud*, vol. 7. London: Hogarth Press, 1905.

Friedmann, John. "The World City Hypothesis." *Development and Change* 17 (1986): 69–83.

Fuchs, Rudi H. *Richard Long*. London: Thames & Hudson, 1986.

Gaar, Gillian G. "Thumping the Nerve Bible: A Conversation with Laurie Anderson." *The Rocket* (Seattle), March 30–April 13, 1994.

Galbraith, John Kenneth. *The New Industrial State*. New York: Houghton Mifflin, 1967.

Gallerneaux, Kristen. *High Static, Dead Lines: Sonic Spectres and the Object Hereafter*. London: Strange Attractor Press, 2018.

Galloway, Alexander R. *The Interface Effect*. Malden, Mass.: Polity, 2013.

Galloway, Alexander R., Eugene Thacker, and McKenzie Wark. *Excommunication: Three Inquiries in Media and Mediation*. Chicago: University of Chicago Press, 2014.

Ganz, Jacob, and Joel Rose. "The MP3: A History of Innovation and Betrayal." *The Record: Music News from NPR*, March 23, 2011. https://www.npr.org/sections/therecord /2011/03/23/134622940/the-mp3-a-history-of-innovation-and-betrayal.

Garreau, Joel. *Edge City: Life on the New Frontier*. New York: Doubleday, 1991.

Gelitz, Christiane. "You Are What You Like." *Scientific American Mind*, March–April 2011, 38–43.

"Generational Insights and the Speed of Change." American Marketing Association, June 30, 2022. https://www.ama.org/marketing-news/generational-insights-and-the-speed-of -change.

Genette, Gérard. *Paratexts: Thresholds of Interpretation*. Cambridge, UK: Cambridge University Press, 1997.

Gentner, Dedre, and Jonathan Grudin. "The Evolution of Mental Metaphors in Psychology: A 90-Year Retrospective." *American Psychologist* 40, no. 2 (February 1985), 181–91.

Gergen, Kenneth. *The Saturated Self: Dilemmas of Identity in Contemporary Life*. New York: Basic Books, 1991.

Germs. "What We Do Is Secret!" In *G.I.* LP. Los Angeles: Slash Records, 1979.

Gibbs-Smith, Charles H. *The Aeroplane: An Historical Survey of Its Origins and Development*. London: HSMO, 1960.

Gibson, James J. *The Ecological Approach to Visual Perception*. New York: Houghton Mifflin, 1979.

———. *The Senses Considered as Perceptual Systems*. New York: Houghton Mifflin, 1966.

Gibson, William. *All Tomorrow's Parties*. New York: Putnam, 1999.

———. "Burning Chrome." *Omni*, July 1982, 72–77, 102–7.

———. *Count Zero*. London: Voyager, 1986.

———. *Distrust That Particular Flavor*. New York: Putnam, 2012.

———. *Idoru*. New York: Putnam, 1996.

———. "Johnny Mnemonic." *Omni*, May 1981, 56–63, 98–99.

———. "The Recombinant City: A Foreword." In *Dhalgren*, by Samuel R. Delany. New York: Vintage, 2001.

———. *Spook Country*. New York: Viking, 2007.

———. *Virtual Light*. New York: Bantam, 1993.

Giedion, Sigfried. *Mechanization Takes Command: A Contribution to Anonymous History*. University of Minnesota Press, 2013.

Giggs, Rebecca. *Fathoms: The World in the Whale*. New York: Simon & Schuster, 2020.

Gilder, George. "Telecosm: Metcalfe's Law and Legacy." *Forbes ASAP*, September 13, 1993, 158.

Gilliam, Terry, dir. *Brazil*. DVD. Universal City, Calif.: Universal Home Video, 1998. Theatrical release: 1985.

Gimpel, Jean. *The Medieval Machine: The Industrial Revolution of the Middle Ages*. New York: Penguin, 1976.

Gitelman, Lisa. *Always Already New: Media, History and the Data of Culture*. Cambridge, Mass.: MIT Press, 2006.

———. *Paper Knowledge: Toward a Media History of Documents*. Durham: Duke University Press, 2014.

———. "Unexpected Pleasures: Phonographs and Cultural Identities in America, 1895–1915." In *Appropriating Technology: Vernacular Science and Social Power*, edited by Ron Eglash, Jennifer L. Croissant, Giovanna Di Chiro, and Rayvon Fouché. Minneapolis: University of Minnesota Press, 2004.

Gitlin, Todd. *Media Unlimited: How the Torrent of Images and Sounds Overwhelms Our Lives*. New York: Metropolitan Books, 2001.

Gladwell, Malcolm. *The Tipping Point: How Little Things Can Make a Big Difference*. New York: Little, Brown, 2000,

Gleick, James, *Chaos: Making a New Science*. New York: Viking, 1987.

———. *The Information: A History, a Theory, a Flood*. New York: Pantheon, 2011.

Goddard, Cliff, and Anna Wierzbicka. "Cultural Scripts: What Are They and What Are They Good For?" *Intercultural Pragmatics* 1, no. 2 (2004): 154.

Gold, Matthew K., ed. *Debates in the Digital Humanities*. Minneapolis: University of Minnesota Press, 2012.

Golding, William. *Lord of the Flies*. New York: Penguin, 1954.

González, Jennifer, Kim Gordon, and Matthew Higgs. *Christian Marclay*. London: Phaidon, 2005.

Goodyear, Sarah. "An Illustrated History of All the Ways Urban Environments Can Control Us: An Interview with Sophie Yanow." *Bloomberg*, March 24, 2014. https://www.bloomberg.com/news/articles/2014-03-24/an-illustrated-history-of-all-the-ways-urban-environments-can-control-us.

Gordinier, Jeff. *X Saves the World: How Generation X Got the Shaft but Can Still Keep Everything from Sucking*. New York: Penguin, 2008.

Gordon, Jeremy. "The Evolution of Steve Albini: 'If the dumbest person is on your side, you're on the wrong side.'" *Guardian*, August 15, 2023.

Gordon, Mel. "Laurie Anderson: Performance Artist." *Drama Review* 24, no. 2 (June 1980), 51–54.

Graham, Dan. "Malcolm McLaren and the Making of Annabella." In *Impresario: Malcolm McLaren and the British New Wave*, edited by Paul Taylor. Cambridge, Mass.: MIT Press, 1988.

Grainge, Paul. *Monochrome Memories: Nostalgia and Style in Retro America*. Westport, Conn.: Praeger, 2002.

Granovetter, Mark S. "The Strength of Weak Ties." *American Journal of Sociology* 78, no. 6 (May 1973): 1360–80.

Gray, John. *Straw Dogs: Thoughts on Humans and Other Animals*. New York: Farrar, Straus, & Giroux, 2002.

Greenfield, Adam. *Against the Smart City*. New York: Do Projects, 2013.

———. *Everyware: The Dawning Age of Ubiquitous Computing*. Berkeley: New Riders, 2006.

Greenwald, Andy. *Nothing Feels Good: Punk Rock, Teenagers, and Emo*. New York: St. Martin's, 2003.

Grodin, Debra, and Thomas R. Lindlof. *Constructing the Self in a Mediated World*. Thousand Oaks, Calif.: Sage, 1996.

Gros, Frédéric. *A Philosophy of Walking*. London: Verso, 2014.

Gross, Jason. "Gang of Four: Andy Gill Interview." *Perfect Sound Forever*, October 2000. https://www.furious.com/perfect/gangoffour.html.

Grubbs, David. *Records Ruin the Landscape*. Durham: Duke University Press, 2014.

Guattari, Félix. *Machinic Eros: Writings on Japan*. Minneapolis: Univocal, 2015.

Haddon, Mimi. *What Is Post-Punk? Genre and Identity in Avant-Garde Popular Music, 1977–82*. Ann Arbor: University of Michigan Press, 2020.

Hagle, Will. "Are You Smarter Than an Algorithm?" *Passion of the Weiss*, April 27, 2023. https://www.passionweiss.com/2023/04/27/shazam-songs-megamix.

Hagood, Mack. *Hush: Media and Sonic Self-Control*. Durham: Duke University Press, 2019.

Hales, Peter Bacon. "Grid, Regulation, Desire Line: Contests over Civic Space in Chicago." In *Public Space and the Ideology of Place in American Culture*, edited by Miles Orville and Jeffrey L. Meikle. New York: Rodopi, 2009.

Halpern, Susan. *Four Wings and a Prayer*. New York: Vintage, 2001.

Hampton, Bruce, and David Cole. *Soft Paths*. Harrisburg, Pa.: Stackpole, 1988.

Harkaway, Nick. *The Blind Giant: How to Survive in the Digital Age*. London: John Murray, 2012.

Harris, Michael. *The End of Absence: Reclaiming What We've Lost in a World of Constant Connection*. New York: Current, 2014.

Harvey, David. *Rebel Cities: From the Right to the City to the Urban Revolution*. New York: Verso, 2012.

Hawk, Tony. "From Carving to Flying." In *California Concrete: A Landscape of Skateparks*, by Amir Zaki. London: Merrell, 2019.

Hayles, N. Katherine. *Chaos Bound: Orderly Disorder in Contemporary Literature and Science*. Ithaca: Cornell University Press, 2018.

———. *How We Became Posthuman: Virtual Bodies in Cybernetics, Literature, and Informatics*. Chicago: University of Chicago Press, 1999

Haynsworth, Leslie. "'Alternative' Music and the Oppositional Potential of Generation X Culture." In *GenXegesis: Essays on Alternative Youth (Sub)Culture,* edited by Ulrich, John M. and Andrea L. Harris. Madison: University of Wisconsin Press, 2003.

Headlam, Bruce. "Origins; Walkman Sounded Bell for Cyberspace." *New York Times*, July 29, 1999.

"Heap of Hype Heralds Sigue Arrival." *Chicago Tribune*, November 6, 1986.

Hegarty, Paul. *Noise/Music: A History*. New York: Bloomsbury, 2008.

———. *Rumour and Radiation: Sound in Video Art*. New York: Bloomsbury, 2015.

Hegarty, Paul, and Martin Halliwell. *Beyond and Before: Progressive Rock Since the 1960s*. New York: Continuum, 2011.

Heim, Michael. *Electric Language: A Philosophical Study of Word Processing*. New Haven: Yale University Press, 1993.

———. *The Metaphysics of Virtual Reality*. New York: Oxford University Press, 1993.

Heraclitus. *Fragments*. New York: Viking, 2001.

Herbert, Frank. *Dune Messiah*. New York: Putnam, 1969.

Herzog, Thomas, and Jennifer A. Flynn-Smith. "Preference and Perceived Danger as a Function of the Perceived Curvature, Length, and Width of Urban Alleys." *Environment and Behavior* 33, no. 5 (September 2001): 653–66.

Heys, Toby. "Muzak and the Working Dead." In *AUDINT: Unsound: Undead*, edited by Steve Goodman, Toby Heys, and Eleni Ikoniadou. Falmouth, UK: Urbanomic, 2019.

Hill, Mike, dir. *The Man Who Souled the World*. DVD. Los Angeles: Whyte House, 2007.

Hoffmann, Susanne, et al. "Duets Recorded in the Wild Reveal That Interindividually Coordinated Motor Control Enables Cooperative Behavior." *Nature Communications*, June 12, 2019, 2577.

Hofstadter, Douglas. *I Am a Strange Loop*. New York: Basic, 2007.

Holston, James. *The Modernist City: An Anthropological Critique of Brasília*. Chicago: University of Chicago Press, 1989.

Home, Stewart. "How I Discovered America." *Info Pool*, no. 6 (2002). https://www.stewart homesociety.org/art/america.htm.

Houston, Alex. "Xerox and Roll: The Corporate Machine and the Making of Punk." *JSTOR Daily*, October 22, 2023. https://daily.jstor.org/xerox-and-roll-the-corporate-machine -and-the-making-of-punk.

Howe, Neil, and Bill Strauss. *13th Gen: Abort, Retry, Ignore, Fail?* New York: Vintage, 1993.

Howell, John. *Laurie Anderson*. American Originals. New York: Thunder's Mouth Press, 1992.

Howell, Linda. "The Cyborg Manifesto Revisited: Issues and Methods for Technocultural Feminism." In *Postmodern Apocalypse: Theory and Cultural Practice at the End*, edited by Richard Dellamora. Philadelphia: University of Pennsylvania Press, 1995.

Howell, Ocean. "The Poetics of Security: Skateboarding, Urban Design, and the New Public Space." UrbanPolicy.net, 2001. https://urbanpolicy.net/wp-content/uploads/2013/02 /Howell_2001_Poetics-of-Security_NoPix.pdf.

"How 130-Year-Old National Geographic Became a Top Social Media Brand." Marketing-Interative, February 22, 2018.

Hughes, June Sananjaleen. *Whale Wisdom, Dolphin Joy: Ascension Teachings from the Cetaceans*. Rectortown, Va.: J. Hughes, 2014.

Huhtamo, Erkki, and Jussi Parikka, eds. *Media Archaeology: Approaches, Applications, and Implications*. Berkeley: University of California Press, 2011.

Hunter, Robert. *The Preservation of Open Spaces, and of Footpaths, and Other Rights of Way: A Practical Treatise on the Law of the Subject*. London: Eyre & Spottiswood, 1896.

Huxley, Aldous. *America and the Future: An Essay*. Austin, Tex.: Jenkins, 1970.

Huyssen, Andreas. *Present Pasts: Urban Palimpsests and the Politics of Memory*. Stanford: Stanford University Press, 2003.

Hyden, Steven. "How Radiohead's 'Kid A' Kicked Off the Streaming Revolution." *Grantland*, September 29, 2015. http://grantland.com/hollywood-prospectus/how-radioheads-kid-a-kicked-off-the-streaming-revolution.

Iacoboni, Marco. *Mirroring People*. New York: Farrar, Straus, and Giroux, 2008.

Ingold, Tim. *The Life of Lines*. New York: Routledge, 2015.

———. *Lines: A Brief History*. New York: Routledge, 2007.

Innis, Harold A. *The Bias of Communication*. Toronto: University of Toronto Press, 1951.

Ishiguro, Kazuo. *Klara and the Sun*. New York: Knopf, 2021.

Ito, Toshiharu. "The Future Form of Visual Art." *Art and Design*, no. 39 (1994): 66–69.

Jackson, John Brickerson. *A Sense of Place, a Sense of Time*. New Haven: Yale University Press, 1996.

Jacobs, Jane. *The Economy of Cities*. New York: Random House, 1969.

Jakab, Peter L. *Visions of a Flying Machine: The Wright Brothers and the Process of Invention*. London: Smithsonian Institution Press, 1990.

Jansen, Bas, "Tape Cassettes and Former Selves: How Mix Tapes Mediate Memories." In *Sound Souvenirs: Audio Technologies: Memory and Cultural Practices*, edited by Karen Bijsterveld and José van Dijck. Amsterdam: Amsterdam University Press, 2009.

Jenkins, Henry. *Convergence Culture: Where Old and New Media Collide*. New York: New York University Press, 2006.

Jeter, K. W. *Noir*. New York: Bantam, 1998.

Johnson, Steven. *The Ghost Map*. New York: Riverhead, 2006.

Jones, Brian Jay. *George Lucas: A Life*. New York: Little, Brown, 2016.

Jonze, Spike. "Steve Rocco Interview." *Dirt Magazine*, no. 3 (1992), 26–27.

Kaczynski, Theodore. *Technological Slavery: The Collected Writings of Theodore J. Kaczynski, a.k.a. "The Unabomber."* Port Townsend, Wash.: Feral House, 2010.

Kagge, Erling. *Walking: One Step at a Time*. New York: Pantheon, 2019.

Kahn, Douglas. "Christian Marclay's Early Years: An Interview." *Leonardo Music Journal* 13 (2003): 17–21.

Katz, Mark. *Capturing Sound: How Technology Has Changed Music*. Berkeley: University of California Press, 2004.

Kaufmann, Will. *The Comedian as Confidence Man: Studies in Irony Fatigue*. Detroit: Wayne State University Press, 1997.

Kelling, George L., and Catherine M. Coles. *Fixing Broken Windows: Restoring Order and Reducing Crime in Our Communities*. New York: Free Press, 1996.

Kellner, Douglas. *Cultural Studies, Identity, and Politics: Between the Modern and the Postmodern*. New York: Routledge, 1995.

Kelly, Kevin. "Mr. Big Trend." *Wired*, October 1994, 114–15.

———. *New Rules for the New Economy: 10 Radical Strategies for a Connected World*. New York: Viking, 1998.

———. *What Technology Wants*. New York: Penguin, 2010.

Kelly, Richard E., dir. *Donnie Darko*. DVD. Beverly Hills: 20th Century Fox Home Entertainment, 2001.

Keyes, Daniel. *Flowers for Algernon*. New York: Harcourt, 1966.

Khazam, Rahma. "Jumpcut Jockey." In *On and By Christian* Marclay, edited by Jean-Pierre Criqui. London: Whitechapel Gallery, 2014.

Kiberd, Roisin. *The Disconnect: A Personal Journey Through the Internet*. London: Serpent's Tail, 2021.

Kilcullen, David. *Out of the Mountains: The Coming Age of the Urban Guerrilla*. New York: Oxford University Press, 2013.

Killer Mike. "Killer Mike Baltimore Op-Ed." *Billboard*, May 1, 2015: https://www.billboard .com/music/music-news/killer-mike-baltimore-op-ed-correspondents-dinner-freddie -gray-6553719.

Kimmerer, Robin Wall. *Braiding Sweetgrass: Indigenous Wisdom, Scientific Knowledge, and the Teachings of Plants*. Minneapolis: Milkweed Editions, 2013.

King, Emily. "Laurie Anderson in Conversation." *The Happy Reader*, no. 12 (Winter 2018), 10–34.

Kittler, Friedrich A. *Discourse Networks: 1800/1900*. Stanford: Stanford University Press, 1990.

———. *Gramophone, Film, Typewriter*. Stanford: Stanford University Press, 1999.

———. *Literature, Media, Information Systems: Essays*. G+B Arts, 1997.

———. *Optical Media*. Cambridge, UK: Polity, 2010.

Kleberg, Lars. *Theatre as Action: Soviet Russian Avant-Garde Aesthetics*. New York: New York University Press, 1993.

Klein, Ezra. "I Didn't Want It to Be True, but the Medium Really Is the Message." *New York Times*, August 7, 2022.

Klein, Norman M. *The History of Forgetting: Los Angeles and the Erasure of Memory*. New York: Verso, 1997.

Kleinman, Zoe. "MySpace Admits Losing 12 Years' Worth of Music Uploads." *BBC News*, March 18, 2019. https://www.bbc.com/news/technology-47610936.

Kluitenberg, Eric. *Delusive Spaces: Essays on Culture, Media and Technology*. Rotterdam: NAi, 2008.

———. "The Society of the Unspectacular." *Autonomedia*, June 10, 2007: http://dev
.autonomedia.org/node/5573.

Knapp, Bettina. *Machine, Metaphor, and the Writer: A Jungian View*. University Park: Pennsylvania State University Press, 1989.

Krims, Adam. *Music and Urban Geography*. New York: Routledge, 2007.

Kroker, Arthur. *Spasm: Virtual Reality, Android Music and Electric Flesh*. New York: St.
Martin's, 1993.

Kugelberg, Johan. Foreword to *I'll Kick You in the Head with My Energy Legs*, by Jonnie
Craig. Årsta, Sweden: Dokument, 2012.

Kumar, Aparna. "Libertarian, or Just Bizarro?" *Wired*, April 25, 2001. https://www.wired
.com/2001/04/libertarian-or-just-bizarro.

Kunzru, Hari. "Attention."" *Harper's*, July 2021. https://harpers.org/archive/2021/07
/attention-hari-kunzru.

Kurland, Andrea. "Getting Deep with Ian MacKaye, the Godfather of DIY Culture." *Huck
Magazine*, no. 49 (2015). https://www.huckmag.com/art-and-culture/music-2/ian
-mackaye-survival-issue-interview.

Kyl C., and W. Gronen. "The Bicycle-Airplane Connection." *Air and Space*, February/March
1990, 77–79.

LaBelle, Brandon. *Acoustic Territories: Sound Culture and Everyday Life*. New York: Continuum, 2010.

Lakoff, George. "The Contemporary Theory of Metaphor." In *Metaphor and Thought*, edited
by Andrew Ortony. Cambridge, UK: Cambridge University Press, 1993.

Lakoff, George, and Mark Johnson. *Metaphors We Live By*. Chicago: University of Chicago
Press, 1980.

La Mettrie, Julien Offray de. *Man, a Machine*. LaSalle, Ill.: Open Court, 1912.

Lanier, Jaron. *Dawn of the New Everything: Encounters with Reality and Virtual Reality*. New
York: Henry Holt, 2017.

———. *Ten Arguments for Deleting Your Social Media Accounts Right Now*. New York: Holt,
2018.

———. "Where Did the Music Go?" In *Sound Unbound: Sampling Digital Music and Culture*, edited by Paul D. Miller. Cambridge, Mass.: MIT Press, 2008.

———. *Who Owns the Future?* New York: Simon & Schuster, 2013.

Lanois, Daniel. *Soul Mining: A Musical Life*. London: Faber & Faber, 2010.

Latour, Bruno. "Beware, Your Imagination Leaves Digital Traces." *Times Higher Education
Supplement*, April 6, 2007.

———. *Pandora's Hope: Essays on the Reality of Science Studies*. Cambridge, Mass.: Harvard
University Press, 1999.

Le Corbusier. *The City of To-morrow and Its Planning*. 3rd ed. London: Architectural Press,
1978.

Leigh, Danny. "'I just kept cool, you know. Travelled. Did a couple of porn movies.'" *Guardian*, February 9, 2001.

LeRoy, Dan, and Peter Relic. *For Whom the Cowbell Tolls: 25 Years of Paul's Boutique*. Akron, Ohio: 6623 Press, 2014.

Leung, Godfre, "Figure and Ground, Image and Sound: The Digital Medium in Brian Eno's *Thursday Afternoon*." *The Soundtrack* 9, nos. 1 and 2 (2016), 91–105.

Levin, Thomas Y. "Indexically Concrète: The Aesthetic Politics of Christian Marclay's Gramophonia." *Parkett* 56 (1999), 162–69.

Levinson, Paul. *Digital McLuhan: A Guide to the Information Millennium*. New York: Routledge, 1999.

Levy, Steven. *The Perfect Thing*. New York: Simon & Schuster, 2006.

Leyton, Michael. *Symmetry, Causality, Mind*. Cambridge, Mass.: MIT Press, 1992.

Li, Yueh-Tuan, and Wen-Shu Lai. "Voice, Object and Listening in the Sound Installations of Laurie Anderson." *International Journal of Performance Arts and Digital Media* 9, no. 2 (2013): 347–61.

Licht, Alan, "CBGB as Imaginary Landscape: The Music of Christian Marclay." In *Cristian Marclay*, edited by Russell Ferguson. Los Angeles: Hammer Museum, 2004.

Lidwell, William, Kritina Holden, and Jill Butler. *Universal Principles of Design*. Beverly, Mass.: Rockport, 2010.

Lienhard, John. *The Engines of Our Ingenuity: An Engineer Looks at Technology and Culture*. New York: Oxford University Press, 2000.

Linge, Nigel. "The Archaeology of Communications' Digital Age." *Industrial Archaeology Review* 35 (May 2013): 45–64.

Lingel, Jessa. *Digital Countercultures and the Struggle for Community*. Cambridge, Mass.: MIT Press, 2017.

Litwin, Rory. *Speaking of Information*. Duluth: Library Juice, 2009.

Liu, Alan. *The Laws of Cool: Knowledge Work and the Culture of Information*. Chicago: University of Chicago Press, 2004.

Lockwood, Patricia. *No One Is Talking About This*. New York: Riverhead Books, 2021.

Logan, Roger K. *McLuhan Misunderstood: Setting the Record Straight*. Toronto: Key Publishing House, 2013.

Long, Richard. *Heaven and Earth*. London: Tate, 2009.

———. *Selected Statements and Interviews*. London: Haunch of Venison, 2007.

———. *Walking the Line*. London: Thames & Hudson, 2002.

Louison, Cole. *The Impossible: Rodney Mullen, Ryan Sheckler, and the Fantastic History of Skateboarding*. Guilford, Conn.: Lyons, 2011.

Lovink, Geert. *Dark Fiber: Tracking Critical Internet Culture*. Cambridge, Mass.: MIT Press, 2003.

———. *My First Recession: Critical Internet Culture in Transition*. Rotterdam: V2/NAi Publishers, 2004.

Lund, Christian. "Laurie Anderson Interview: A Life of Storytelling." *Louisiana Channel*, May 2016. https://www.youtube.com/watch?v=dUo-dqMriY8.

Lynch, Kevin. *The Image of the City*. Cambridge, Mass.: MIT Press, 1960.

———. *Managing the Sense of a Region*. Cambridge, Mass.: MIT Press, 1976.

———. *What Time Is This Place?* Cambridge, Mass.: MIT Press, 1972.

Lynch, Kevin, and Gary Hack. *Site Planning*. Cambridge, Mass.: MIT Press, 1984.

Ma, Shouwen, Andries ter Maat, and Manfred Gahr. "Neurotelemetry Reveals Putative Predictive Activity in HVC During Call-Based Vocal Communications in Zebra Finches." *Journal of Neuroscience* 40, no. 32 (August 5, 2020): 6219–27.

Mac Cormac, Earl R. *A Cognitive Theory of Metaphor*. Cambridge, Mass.: MIT Press, 1985.

Madera, Gerardo. *Name, Thing, Thing: A Primer in Parallel Typographies*. New York: Printed Matter, 2018.

Mailander, Joseph F. *LA at Intermission: A City Mingling Towards Identity*. Los Angeles: Nellcôte Press, 2014.

Malhi, Gin S. "Quality of Life. . . on Earth." *Acta Neuropsychiatrica*, 2010, 22.

Marchessault, Janet. *Ecstatic Worlds: Media, Utopias, Ecologies*. Cambridge, Mass.: MIT Press, 2017.

Marchland, Philip. *Marshall McLuhan: The Medium and the Messenger*. Boston: Ticknor & Fields, 1989.

Marclay, Christian. *Christian Marclay*. Exhibition catalog. Edited by Russell Ferguson. Los Angeles: Steidl, 2003.

———. *Record Without a Cover*. New York: Recycled Records, 1985.

Marcus, Ben. *The Skateboard: The Good, the Rad, and the Gnarly*. Minneapolis: MVP, 2011.

Marcus, Gary F. *Kluge: The Haphazard Construction of the Human Mind*. New York: Houghton Mifflin, 2008.

Marcus, Greil. "Heading for the Hills." *East Bay Express*, February 19, 1999.

———. *Lipstick Traces: A Secret History of the Twentieth Century*. Cambridge, Mass.: Harvard University Press, 1989.

Mars, Roman, and Kurt Kohlstedt. *The 99% Invisible City: A Field Guide to the Hidden World of Everyday Design*. New York: Houghton Mifflin, 2020.

Martin, Graham Dunstan. *Shadows in the Cave: Mapping the Conscious Universe*. New York: Arkane, 1990.

Massumi, Brian. *A User's Guide to Capitalism and Schizophrenia: Deviations from Deleuze and Guattari*. Cambridge, Mass.: MIT Press, 1992.

Mattern, Shannon. *Code+Clay . . . Data+Dirt: Five Thousand Years of Urban Media*. Minneapolis: University of Minnesota Press, 2017.

———. "Sidewalks of Concrete and Code." In *Re-Understanding Media: Feminist Extensions of Marshall McLuhan*, edited by Sarah Sharma and Rianka Singh. Durham: Duke University Press, 2022.

Maturana, Humberto, and Bernhard Poerkson. *From Being to Doing: The Origins of the Biology of Cognition*. Heidelberg, Germany: Carl-Auer Verlag, 2004.

Maturana, Humberto, and Francisco J. Varela. *The Tree of Life: The Biological Roots of Human Understanding*. Boston: Shambhala, 1987.

Maughan, Tim. *Infinite Detail*. New York: Farrar, Straus and Giroux, 2019.

McCarthy, Cormac. *The Passenger*. New York: Knopf, 2022.

———. *Stella Maris*. New York: Knopf, 2022.

McCloud, Scott. *Understanding Comics: The Invisible Art*. New York: HarperCollins, 1993.

McCorduck, Pamela. "America's Multi-Mediatrix." *Wired*, March 1994, https://www.wired.com/1994/03/anderson.

McCullough, Malcolm. *Ambient Commons: Attention in the Age of Embodied Information*. Cambridge, Mass.: MIT Press, 2013.

McDonough, Tom. "Situationist Spaces." In *Guy Debord and the Situationist International: Texts and Documents*, edited by Tom McDonough. Cambridge, Mass.: MIT Press, 2002.

McGee, Alan. *Creation Stories: Riots, Raves, and Running a Label*. London: Pan, 2013.

McGlone, Matthew S., Gary Beck, and Abigail Pfiester. "Contamination and Camouflage in Euphemisms." *Communication Monographs* 73, no. 3 (September 2006): 261–82.

McKenna, Terence. *The Archaic Revival: Speculations on Psychedelic Mushrooms, the Amazon, Virtual Reality, UFOs, Evolution, Shamanism, the Rebirth of the Goddess, and the End of History*. San Francisco: Harper San Francisco, 1992.

McLaren, Malcolm, ed. *Musical Paintings*. Zurich: JPR/Ringier, 2009.

McLuhan, Marshall. *Counterblast*. London: Rapp & Whiting, 1969.

———. *Culture Is Our Business*. New York: Ballantine, 1970.

———. *The Gutenberg Galaxy: The Making of Typographical Man*. Toronto: University of Toronto Press, 1962.

———. *Understanding Media: The Extensions of Man*. New York: Houghton Mifflin, 1964.

McLuhan, Marshall, and Quintin Fiore. *War and Peace in the Global Village*. New York: Bantam, 1968.

McLuhan, Marshall, and Eric McLuhan. *Laws of Media: The New Science*. Toronto: University of Toronto Press, 1988.

———. *The Lost Tetrads of Marshall McLuhan*. New York: O/R, 2017.

———. *Media and Formal Cause*. Houston: NeoPoiesis Press, 2011.

McMahon, C. J., Jr., and Chad D. Graham Jr. *Introduction to Engineering Materials: The Bicycle and the Walkman*. Philadelphia: Merion, 1992.

Means, James. "Wheeling and Flying." *Aeronautical Annual*, 1896, 24–25.

Melton, Jeffrey, and Eric Sterling. "The Subversion of Happy Endings in Terry Gilliam's *Brazil*." In *The Cinema of Terry Gilliam: It's a Mad World*, edited by Jeff Birkenstein, Anna Froula, and Karen Randell. New York: Wallflower, 2013.

Merholz, Peter. "Ethnoclassification and Vernacular Vocabularies." PeterMe, August 30, 2004. https://www.peterme.com/archives/000387.html

———. "Metadata for the Masses." *Adaptive Path*, October 19, 2004. http://www
.adaptivepath.com/publications/essays/archives/000361.php.

Meyrowitz, Joshua. "Medium Theory." In *Communication Theory Today*, edited by David
Crowley and David Mitchell. Stanford: Stanford University Press, 1994.

———. *No Sense of Place: The Impact of Electronic Media on Social Behavior*. New York:
Oxford University Press, 1985.

Michell, John. Introduction to *The Old Straight Track*, by Alfred Watkins. London: Garn-
stone, 1970.

———. *The New View over Atlantis*. New York: Thames & Hudson, 1995.

Milemarker. "Frigid Forms Sell You Warmth." In *Frigid Forms Sell*. LP. Arlington, Va.: Lovitt
Records, 2000.

Milutis, Joe. *Ether: The Nothing That Connects Everything*. Minneapolis: University of Minnesota Press, 2005.

Miroshnichenko, Andrey. *Human as Media: The Emancipation of Authorship*. A. Miroshnichenko, 2020.

Mitchell, Melanie, *Complexity: A Guided Tour*. New York: Oxford University Press, 2009.

Molinaro, Mattie, Corinne McLuhan, and William Toye, eds. *Letters of Marshall McLuhan*.
New York: Oxford University Press, 1987.

Momus. "Pop Stars? Nein Danke!" 1991. https://imomus.com/index499.html.

Montgomery, Charles. *Happy City: Transforming Our Lives Through Urban Design*. New
York: Farrar, Straus & Giroux, 2013.

Moody, Rick. Foreword to *How to Write About Music*, edited by Marc Woodworth and
Ally-Jane Grossan. New York: Bloomsbury, 2015.

Mooney, Ted. *Easy Travel to Other Planets*. New York: Farrar Straus & Giroux, 1981.

Moore, Catherine. "Works and Recordings: The Impact of Commercialism and Digitalisa-
tion." In *The Musical Work: Reality or Invention?*, edited by Michael Talbot. Liverpool:
Liverpool University Press, 2000.

Moore, Thurston. *Mix Tape: The Art of Cassette Culture*. New York: Universe, 2004.

Morley, Paul. "On Gospel, Abba, and the Death of the Record: An Audience with Brian
Eno." *Guardian*, January 16, 2010.

Morrissey and Johnny Marr. "How Soon Is Now?" In *Hatful of Hollow*, by the Smiths. LP.
London: Rough Trade, 1984.

Morrison, Grant. *The Invisibles* 1, no. 3 (November 1994).

Morrison, Tony. *Pathways to the Gods*. New York: Harper & Row, 1978.

Mortimer, Sean. *Stalefish: Skateboard Culture from the Rejects Who Made It*. San Francisco:
Chronicle, 2008.

Morton, Timothy. *The Ecological Thought*. Cambridge, Mass.: Harvard University Press, 2012.

Morville, Peter. *Ambient Findability: What We Find Changes Who We Become*. Sebastopol,
Calif.: O'Reilly Media, 2005.

Munzenrider, Mike. "Starting a New Skateboard Magazine and Other Radical Acts of Love."

Quartersnacks, June 28, 2023. https://quartersnacks.com/2023/06/starting-a-new -skateboard-magazine-and-other-radical-acts-of-love.

Murphy, Douglas. *Last Futures: Nature, Technology, and the End of Architecture*. New York: Verso, 2016.

Muybridge, Eadweard. *Animals in Motion*. New York: Dover, 1957.

Myhill, Carl. "Commercial Success by Looking for Desire Lines." In *Computer-Human Interaction: 6th Asia Pacific Conference, APCHI 2004*, edited by Masood Masoodian, Steve Jones, and Bill Rogers. Berlin: Springer-Verlag, 2004.

Naisbitt, John, Nana Naisbitt, and Douglas Philips. *High Tech/High Touch: Technology and Our Search for Meaning*. New York: Broadway Books, 1999.

Nash, Jørgen, Jens Jørgen Thorsen, and Dieter Kunzelmann. "Slogans." In *Cosmonauts of the Future: Texts from the Situationist Movement in Scandinavia and Elsewhere*, edited Mikkel Bolt Rasmussen and Jakob Jakobsen. Copenhagen: Nebula, 2015.

Neale, Mark, dir. *William Gibson: No Maps for These Territories*. Video. New York: Docurama, 2000.

Nedorostek, Nathan, and Anthony Pappalardo. *Radio Silence: A Selected Visual History of American Hardcore Music*. New York: MTV Press, 2008.

Needs, Kris. "Sigue Sigue Sputnik: *Flaunt It*." *Record Collector*, no. 513 (December 25, 2020).

Negarestani, Reza. *Cyclonopedia: Complicity with Anonymous Materials*. Melbourne, Australia: re.press, 2008.

Negroponte, Nicholas. *Being Digital*. New York: Knopf, 1995.

Neill, Ben. "Christian Marclay." *Bomb*, Summer 2003, 44–51.

Nelson, Theodor Holm. *Literary Machines*. Edition 87.1. South Bend, Ind.: T. H. Nelson, 1987.

———. *Computer Lib/Dream Machines*. Redmond, Wash.: Tempus Books/Microsoft Press, 1987.

Newport, Cal. *Digital Minimalism: Choosing a Focused Life in a Noisy World*. New York: Portfolio, 2019.

Nicholson, Geoff. *The Lost Art of Walking*. New York: Riverhead, 2008.

Nietzsche, Friedrich. *Untimely Meditations*. Cambridge, UK: Cambridge University Press, 1997.

———. "Writing Ball is a thing like me. . ." Typescript, 1882. Cat. no. GSA 71/234, Friedrich Nietzsche estate (item no. 71) at the Goethe- und Schiller-Archiv (GSA), Weimar, Germany.

Nolan, David. *I Swear I Was There: Sex Pistols, Manchester and the Gig That Changed the World*. London: John Blake Publishing, 2016.

The Nonce. "Mixtapes." In *World Ultimate*. LP. New York: American Recordings, 1995.

Norman, Donald. *Living with Complexity*. Cambridge, Mass.: MIT Press, 2010.

———. *Turn Signals Are the Facial Expressions of Automobiles*. Boston: Addison-Wesley, 1992.

Norton, Richard J. "Feral Cities." *Naval War College Review* 56, no. 4 (Autumn 2003): 97–106.

Novak, Matt. "How the 1920s Thought Electricity Would Transform Farms Forever." *Paleofuture*, June 3, 2013. https://paleofuture.com/blog/2013/6/3/how-the-1920s-thought-electricity-would-transform-farms-forever.

———. "Musicians Wage War Against Evil Robots." *Smithsonian Magazine*, February 10, 2012, http://blogs.smithsonianmag.com/paleofuture/2012/02/musicians-wage-war-against-evil-robots.

Oakes, Kaya, *Slanted and Enchanted: The Evolution of Indie Culture.* New York: Holt, 2009.

O'Connor, M. R. *Wayfinding: The Science and Mystery of How Humans Navigate the World.* New York: St. Martin's, 2019.

O'Connor, Stuart. "William Gibson: 'I'm agnostic about technology. But I want a robotic penguin.'" *Guardian*, August 29, 2010.

O'Haver, Hanson. "An Oral History of 'Transworld Skateboarding' Magazine." *Vice*, May 30, 2019. https://www.vice.com/en/article/nea4vd/an-oral-history-of-transworld-skateboarding-magazine.

Ong, Walter J. *Interfaces of the Word: Studies in the Evolution of Consciousness and Culture.* Ithaca: Cornell University Press, 1977.

———. *Orality and Literacy: The Technologizing of the Word.* New York: Routledge, 1982

———. *Rhetoric, Romance, and Technology: Studies in the Interaction of Expression and Culture.* Ithaca: Cornell University Press, 1971.

O'Rourke, Karen. *Walking and Mapping: Artists as Cartographers.* Cambridge, Mass.: MIT Press, 2013.

Open Mike Eagle. "El-P Tells the Story of Putting Out the Funcrusher EP Independently." *What Had Happened Was.* 2021. YouTube, https://www.youtube.com/watch?v=dgJPhZuJ1FY&list=PLDIU8Sm4Gvx9nw2lLosgis16XymDjdW8g.

Ortega y Gasset, José. *The Dehumanization of Art and Ideas About the Novel.* Princeton: Princeton University Press, 1925.

Paine, Jake. "Killer Mike and El-P 'Run the Jewels' Release Date, Tracklist and Album Download." *HipHop DX*, June 26, 2013. https://web.archive.org/web/20140222012008/http://www.hiphopdx.com/index/news/id.23535/title.killer-mike-el-p-run-the-jewels-release-date-tracklist-album-download.

Palahniuk, Chuck. *Fight Club.* New York: Norton, 1996.

———. *Rant: An Oral Biography of Buster Casey.* New York: Doubleday, 2007.

Papacharissi, Zizi. *A Networked Self: Identity, Community, and Culture on Social Network Sites.* New York: Routledge, 2011.

Papanek, Victor. *Design for the Real World: Human Ecology and Social Change.* New York: Pantheon, 1972.

Parikka, Jussi. *The Anthrobscene.* Minneapolis: University of Minnesota Press, 2014.

———. *A Geology of Media.* Minneapolis: University of Minnesota Press, 2015.

———. *Insect Media: An Archeology of Animals and Technology.* Minneapolis: University of Minnesota Press, 2010.

———. *What Is Media Archaeology?* Cambridge, UK: Polity, 2012.

Parmentier, Richard J. *The Sacred Remains: Myth, History, and Polity in Belau*. Chicago: University of Chicago Press, 1987.

Parry, Kyle. *A Theory of Assembly: From Museums to Memes*. Minneapolis: University of Minnesota Press, 2022.

Parser, Eli. *The Filter Bubble: What the Internet Is Hiding from You*. New York: Penguin, 2011.

Patterson, Andrew, dir. *The Vast of Night*. Motion picture. Culver City, Calif.: Amazon Studios, 2019.

Paul, Gregory S. *Dinosaurs of the Air: The Evolution and Loss of Flight in Dinosaurs and Birds*. Baltimore: Johns Hopkins University Press, 2002.

Pawley, Martin. "Architecture Versus the Movies, or Forum Versus Content." *Architectural Design*, no. 6 (June 1970), 26–53.

Pearce, Sheldon. "Considering the Rise of the Super Short Rap Song." *Pitchfork*, March 15, 2018. https://pitchfork.com/thepitch/considering-the-rise-of-the-super-short-rap-song.

Peele, Jordan, dir. *Nope*. Motion picture. Los Angeles: Monkey Paw Productions, 2022.

Peralta, Stacy, dir. *Bones Brigade: An Autobiography*. Motion picture. Santa Monica: Nonfiction Unlimited, 2012.

Perry, David B. *Bike Cult: The Ultimate Guide to Human-Powered Vehicles*. New York: Four Walls Eight Windows, 1995.

Pescovitz, David. "Douglas Rushkoff: *Present Shock*, the Boing Boing interview." *Boing Boing*, April 2, 2013. https://boingboing.net/2013/04/02/douglas-rushkoff-present-shoc.html.

Peters, John Durham. *The Marvelous Clouds: Toward a Philosophy of Elemental Media*. Chicago: University of Chicago Press, 2015.

Pettman, Dominic. *Human Error: Species-Being and Media Machines*. Minneapolis: University of Minnesota Press, 2011.

———. *In Divisible Cities: A Phanto-Cartographical Missive*. Brooklyn: Dead Letter Office, 2013.

———. *Infinite Distraction: Paying Attention to Social Media*. Malden, Mass.: Polity, 2016.

———. *Look at the Bunny: Totem, Taboo, Technology*. London: Zero Books, 2013.

P. H. "Undriven Snow." *Economist*, February 13, 2014. https://www.economist.com/babbage/2014/02/13/undriven-snow.

Pinsker, Joe. "What Will Happen to My Music Library When Spotify Dies?" *Atlantic*, July 19, 2021. https://www.theatlantic.com/culture/archive/2021/07/spotify-streaming-music-library/619453.

Popper, Karl. *Objective Knowledge: An Evolutionary Approach*. Oxford, UK: Oxford University Press, 1972.

Postman, Neil. *Amusing Ourselves to Death: Public Discourse in the Age of Show Business*. New York: Penguin, 1985.

———. "The Reformed English Curriculum." In *High School 1980: The Shape of the Future in American Secondary Education*, edited by Alvin C. Eurich. London: Pitman, 1970.

———. *Technopoly: The Surrender of Culture to Technology*. New York: Knopf, 1992.

Potter, Russell A. *Spectacular Vernaculars: Hip-Hop and the Politics of Postmodernism*. Albany: State University of New York Press,1995.

Power, Nina. "Why Gen X Failed." *Compact*, October 21, 2022. https://compactmag.com /article/why-gen-x-failed.

Prendergast, Mark. *The Ambient Century: From Mahler to Trance—the Evolution of Sound in the Electronic Age*. New York: Bloomsbury, 2000.

Prescott-Steed, David. "Frostbite on My Feet: Representations of Walking in Black Metal Visual Culture." *Helvete: A Journal of Black Metal Theory*, no. 1 (Winter 2013): 45–68.

Prinz, Jessica, *Art Discourse/Discourse in Art*. New Brunswick: Rutgers University Press, 1991.

Purdy, Jedediah. *For Common Things: Irony, Trust, and Commitment in America Today*. New York: Vintage, 1999.

Pynchon, Thomas. *Vineland*. New York: Little, Brown, 1990.

Rabin, Nathan. *You Don't Know Me but You Don't Like Me*. New York: Scribner, 2013.

Raley, Rita. *Tactical Media*. Minneapolis: University of Minnesota Press, 2009.

———. "TXTual Practice." In *Comparative Textual Media: Transforming the Humanities in the Postprint Era*, edited by N. Katerine Hayles and Jessica Pressman. Minneapolis: University of Minnesota Press, 2013.

Ralón, Laureano. "Interview with Eric McLuhan." *Figure/Ground*, August 1, 2010. https:// figureground.org/interview-with-eric-mcluhan.

Rapaport, Herman. *Between the Sign and the Gaze*. Ithaca: Cornell University Press, 1994.

———. "'Can You Say Hello?': Laurie Anderson's *United States*." *Theatre Journal* 38, no. 3 (October 1986): 339–54.

Rapoport, Amos. *The Meaning of the Built Environment: A Nonverbal Communication Approach*. Tucson: University of Arizona Press, 1982.

Rasmussen, Terje. *Communication Technologies and the Mediation of Social Life: Elements of a Social Theory of the New*. Oslo: Department of Media and Communication, University of Oslo, 1996.

———. *Personal Media and Everyday Life*. New York: Springer, 2014.

Raunig, Gerald. *Dividuum: Machinic Capitalism and Molecular Revolution*. Pasadena, Calif.: Semiotext(e), 2016.

———. *A Thousand Machines: A Concise Philosophy of the Machine as Social Movement*. Los Angeles: Semiotext(e), 2010.

Reality Bites: Retirement Anxieties Grow as Generation X Turns 60. Boston: Natixis Investment Managers, 2024. https://www.im.natixis.com/en-us/insights/investor-sentiment /2024/gen-x-report.

Reed, S. Alexander. *Laurie Anderson's "Big Science."* New York: Oxford University Press, 2022.

Reynolds, Simon. *Retromania: Pop Culture's Addiction to Its Own Past*. New York: Faber & Faber, 2011.

———. *Rip It Up and Start Again: Post-Punk, 1978–1984*. London: Faber & Faber, 2005.

Rheingold, Howard. *Net Smart: How to Thrive Online*. Cambridge, Mass.: MIT Press, 2012.

Rhodes, Carl, and Robert Westwood. *Critical Representations of Work and Organization in Popular Culture*. London: Routledge, 2008.

Rhodes, Richard, *Visions of Technology*. New York: Touchstone, 1999.

RIAA (Recording Industry Association of America). "U.S. Recorded Music Revenues by Format, 1973 to 2020." https://www.riaa.com/u-s-sales-database.

Ricoeur, Paul. *Memory, History, Forgetting*. Chicago: University of Chicago Press, 2004.

Richter, Felix. "The Rise and Fall of the Compact Disc." Statista, August 17, 2022. https://www.statista.com/chart/12950/cd-sales-in-the-us.

Riley, Chris. *After the Mass-Age*. Portland, Ore: Analog, 2017.

Rizzolatti, Giacomo, and Corrado Sinigaglia. *Mirrors in the Brain*. New York: Oxford University Press, 2008.

Robertson, Benjamin J. *None of This Is Normal: The Fiction of Jeff VanderMeer*. Minneapolis: University of Minnesota Press, 2018.

Robinson, Kim Stanley. *2312*. New York: Orbit, 2012.

Roelstraete, Dieter. *Richard Long: A Line Made by Walking*. London: Afterall, 2010.

Rogers, Elizabeth Barlow. *Rebuilding Central Park: A Management and Restoration Plan*. Cambridge, Mass.: MIT Press, 1987.

Rombes, Nicholas. *A Cultural History of Punk, 1974–1982*. New York, Continuum, 2009.

Rose, Frank. "The Civil War Inside Sony." *Wired*, February 2003, 100–103, 136–37.

Rose, Mark. *Interstate: Express Highway Politics, 1939–1989*. Knoxville: University of Tennessee Press, 1990.

Rosen, Jody. "The Day the Music Burned." *New York Times*, June 11, 2019.

Rosenberg, Nathan. *Inside the Black Box: Technology and Economics*. Cambridge, UK: Cambridge University Press, 1982.

Rosenberger, Robert. *Callous Objects: Designs Against the Homeless*. Minneapolis: University of Minnesota Press, 2017.

Ross, Alex. "How Björk Broke the Sound Barrier." *Guardian*, February 2015.

Rothenberg, David. *Bug Music: How Insects Gave Us Rhythm and Noise*. New York: Picador, 2013.

Rothman, Wilson. "The Blank Generation: 1979 as Audio Cassette Enabler." *Gizmodo*, July 15, 2009. https://gizmodo.com/the-blank-generation-1979-as-audio-cassette-enabler-5314877.

Rotman, Brian. *Becoming Beside Ourselves: The Alphabet, Ghosts, and Distributed Human Being*. Durham: Duke University Press, 2008.

Rush. "Far Cry." In *Snakes and Arrows*. LP. Toronto: Anthem, 2007.

Rushkoff, Douglas. *Life, Inc.: How Corporatism Conquered the Word, and How We Can Take It Back*. New York: Random House, 2009.

———. *Playing the Future: What We Can Learn from Digital Kids*. New York: HarperCollins, 1996.

———. *Present Shock: When Everything Happens Now*. New York: Current, 2013.

———. *Survival of the Richest: Escape Fantasies of the Tech Billionaires*. New York: Norton, 2022.

———. *Team Human*. New York: Norton, 2019.

Sacks, Mike. *Poking a Dead Frog: Conversations with Today's Top Comedy Writers*. New York: Penguin, 2014.

Sampson, Robert J., and Steve W. Raudenbush. "Systematic Social Observation of Public Spaces: A New Look at Disorder in Urban Neighborhoods." *American Journal of Sociology* 105, no. 3 (November 1, 1999): 603–51.

Sargeant, Will. *Bunnyman: Post-War Kid to Post-Punk Guitarist of Echo and the Bunnymen*. Nashville: Third Man, 2021.

Sassen, Saskia. *The Global City: New York, London, Tokyo*. Princeton: Princeton University Press, 1991.

Saval, Nikil. "Wall of Sound: The iPod Has Changed the Way We Listen to Music. And the Way We Respond to It." *Slate*, March 28, 2011. https://slate.com/culture/2011/03/the -ipod-has-changed-the-way-we-listen-to-music-and-the-way-we-respond-to-it.html.

Schaller, D. "Wing Evolution." In *The Beginnings of Birds: Proceedings of the International Archaeopteryx Conference, Eichstätt, 1984*, edited by Max K. Hecht, J. H. Ostrom, G. Viohl, and P. Wellnhofer. Eichstätt, Germany: Brönner & Daentler, 1985.

Schandorf, Michael. *Communication as Gesture*. Bingley, UK: Emerald, 2019.

Schwartz, Hillel. *The Culture of the Copy: Striking Likenesses, Unreasonable Facsimiles*. New York: Zone, 1996.

Scoates, Christopher. "Perfume, Defense, and David Bowie's Wedding." In *Brian Eno: Visual Music*. San Francisco: Chronicles, 2013.

Sconce, Jeffrey. *Haunted Media: Electronic Presence from Telegraphy to Television*. Durham: Duke University Press, 2000.

Scott, James C. *Domination and the Arts of Resistance: Hidden Transcripts*. New Haven: Yale University Press, 1990.

———. *Seeing Like a State: How Certain Schemes to Improve the Human Condition Have Failed*. New Haven: Yale University Press, 1998.

Scott, Kevin Conroy. "Asking Cosmic Questions: Richard Kelly Interviewed by Kevin Conroy Scott." In Richard Kelly, *The Donnie Darko Book*. London: Faber & Faber, 2003.

Seabrook, John. *Nobrow: The Marketing of Culture and the Culture of Marketing*. New York: Knopf, 2000.

Seymour, Anne. Foreword to *Walking the Line*, by Richard Long. London: Thames & Hudson, 2002.

Shadyac, Tom, dir. *Bruce Almighty*. Motion picture. Los Angeles: Universal Pictures, 2003.

Shapiro, Peter. "Deck Wreckers: The Turntable as Instrument." In *Undercurrents: The Hidden Wiring of Modern Music*, edited by Rob Young. New York: Continuum, 2002.

Shaviro, Steven. *Connected, or What It Means to Live in the Network Society*. Minneapolis: University of Minnesota Press, 2003.

———. *Doom Patrols: A Theoretical Fiction about Postmodernism*. New York: Serpent's Tail, 1996.

Sheffield, Rob. *Love Is a Mix Tape: Life and Loss, One Song at a Time*. New York: Crown, 2007.

Shepard, Nick, and Noëleen Murray. "Introduction: Space, Memory and Identity in the Post-apartheid City." In *Desire Lines: Space, Memory and Identity in the Post-Apartheid City*, by Noëleen Murray, Nick Shepard, and Martin Hall. New York: Routledge, 2007.

Shipman, Pat. *Taking Wing: Archaeopteryx and the Evolution of Bird Flight*. New York: Simon & Schuster, 1998.

Shotter, John. *Conversational Realities: Constructing Life Through Language*. London: Sage, 1993.

Singer, Matthew. "Hotseat: Dave Allen." *Willamette Week*, July 1, 2014. https://www.wweek.com/portland/article-22743-hotseat-dave-allen.html.

Sinker, Daniel, ed. *We Owe You Nothing: Punk Planet, the Collected Interviews*. New York: Akashic, 2001.

Sirius, R. U. "The Importance of Being Andy." *Mondo 2000*, Fall 1989, 84–87.

Smith, Carl. *The Plan of Chicago*. Chicago: University of Chicago Press, 2006.

Smith, Naomi, and Peter Walters. "Desire Lines and Defensive Architecture in Modern Urban Environments." *Urban Studies* 55, no. 13 (2018): 2980–95.

Smith, Patrick A. *Conversations with William Gibson*. Oxford: University Press of Mississippi, 2014.

Solnit, Rebecca. *River of Shadows: Eadweard Muybridge and the Technological Wild West*. New York: Viking, 2003.

———. *Wanderlust: A History of Walking*. New York: Penguin, 2001.

Sommer, Robert. *Personal Space: The Behavioral Basis of Design*. Hoboken, N.J.: Prentice Hall, 1969.

Spigel, Lynn. *Make Room for TV: Television and the Family Ideal in Postwar America*. Chicago: University of Chicago Press, 1992

Spufford, Francis, and Jenny Uglow. *Cultural Babbage: Technology, Time, and Invention*. London: Faber & Faber, 1996.

Star, Susan Leigh. "Slouching Toward Infrastructure." Graduate School of Library and Information Science, University of Illinois, [1996?]. Accessed November 9, 2024, https://web.archive.org/web/20010708162330/http://is.gseis.ucla.edu/research/dl/star.html

Star, Susan Leigh, and James R. Griesemer. "Institutional Ecology, 'Translations' and Boundary Objects: Amateurs and Professionals in Berkeley's Museum of Vertebrate Zoology, 1907–39." *Social Studies of Science* 19, no. 3 (August 1989), 387–420.

Stephenson, Neal. *In the Beginning Was the Command Line*. New York: William Morrow, 1999.

Sterling, Bruce. "Bicycle Repairman." In *Rewired: The Post-Cyberpunk Anthology*, edited by James Patrick Kelly and John Kessell. San Francisco: Tachyon, 2007.

———. "Bruce Sterling on the Art of Text-to-Image Generative AI." Lecture at AI for All,

From the Dark Side to the Light conference, Eindhoven, Netherlands, November 25, 2022, transcribed by Geert Lovink. https://networkcultures.org/blog/2023/05/17/bruce -sterling-on-the-art-of-text-to-image-generative-ai.

———. "October 14, 1998, Beyond the Beyond: Yerba Buena Center for the Arts, San Francisco." *Wired.com*, November 14, 2019. https://www.wired.com/beyond-the-beyond /2019/11/october-14-1998-yerba-buena-center-arts-san-francisco.

Sterling, Bruce, and Richard Kadrey. "Dead Media Manifesto." In *Digital Manifesto Archive*. https://www.digitalmanifesto.net/manifestos/24.

Sterne, Jonathan. *The Audible Past: Cultural Origins of Sound Reproduction.* Durham: Duke University Press, 2003.

Strate, Lance. *Amazing Ourselves to Death: Neil Postman's Brave New World Revisited.* New York: Peter Lang, 2014.

Strauss, William, and Neil Howe. *The Fourth Turning: An American Prophecy.* New York: Three Rivers Press, 1997.

Strode, Muriel. *My Little Book of Prayer.* Chicago: Open Court Press, 1905.

Strogatz, Steven. *Sync: The Emerging Science of Spontaneous Order.* New York: Hyperion, 2003.

Sullivan, Danny. *Ley Lines: The Great Landscape Mystery.* Somerset, UK: Green Magic, 2004.

Swigart, Robert. "A Writer's Desktop." In *The Art of Human-Computer Interface Design*, edited by Brenda Laurel. Reading, Mass.: Addison-Wesley Professional, 1990.

Szendy, Peter. "Christian Marclay on the Phone." In *RE:Play*. Zurich: jrp/ringier, 2007.

Taraska, Julie. "Jewel Personality." *The Wire* (London), June 1995, 18–19.

Taylor, Mark C., Amanda Sharp, and Matthew Higgs. *PressPLAY: Contemporary Artists in Conversation.* London: Phaidon, 2005.

Tenner, Edward. *Our Own Devices: The Past and Future of Body Technology.* New York: Knopf, 2003.

Theil, Philip. *People, Paths, and Purposes: Notations for a Participatory Envirotecture.* Seattle: University of Washington Press, 1997.

Thompson, Gary. "Reweaving the Fabric of the Internet to Transform Humanity." TEDxAustin, February 19, 2011. https://www.youtube.com/watch?v=afMjZgvtsp8.

Thompson, Tade. *Rosewater.* Lexington, Ky.: Apex, 2016.

Thomson, J. Michael. *Great Cities and Their Traffic.* New York: Penguin, 1977.

Throgmorton, James A., and Barbara Eckstein. "Desire Lines: The Chicago Area Transportation Study and the Paradox of Self in Post-War America." [2000?] Accessed November 9, 2024, https://web.archive.org/web/20020414192747/http://www.nottingham.ac.uk /3cities/throgeck.htm.

Tilley, Christopher. *A Phenomenology of Landscape: Places, Paths, and Monuments.* Oxford, UK: Berg, 1997.

Toop, David. "Christian Marclay: Painting in Slang." *The Wire* (London), October 2011, 40–47.

———. *Haunted Weather: Music, Silence, and Memory.* San Francisco: Serpent's Tail, 2004.

———. *Ocean of Sound: Aether Talk, Ambient Sound, and Imaginary Worlds*. San Francisco: Serpent's Tail, 1995.

Torlasco, Domietta. *The Heretical Archive: Digital Memory at the End of Film*. Minneapolis: University of Minnesota Press, 2013.

Toulas, Bill. "Google Drive Users Angry over Losing Months of Stored Data." *Bleeping Computer*, November 27, 2023. https://www.bleepingcomputer.com/news/google/google -drive-users-angry-over-losing-months-of-stored-data.

Tuhus-Dubrow, Rebecca. *Personal Stereo*. New York: Bloomsbury Academic, 2017.

Turchi, Peter. *Maps of the Imagination: The Writer as Cartographer*. San Antonio: Trinity University Press, 2004.

Turkle, Sherry. *Alone Together: Why We Expect More from Technology and Less from Each Other*. New York: Basic Books, 2011.

———. *Life on the Screen: Identity in the Age of the Internet*. New York: Simon & Schuster, 1995.

———. *Reclaiming Conversation: The Power of Talk in a Digital Age*. New York: Penguin, 2015.

Turner, Victor. *The Ritual Process: Structure and Anti-Structure*. Ithaca: Cornell University Press, 1969.

Ullman, Ellen. *Close to the Machine: Technophilia and Its Discontents*. New York: Picador, 1997.

Understanding McLuhan. Interactive CD-ROM. Voyager Interactive, 1996.

Urban, Greg. *Metaculture: How Culture Moves Through the World*. Minneapolis: University of Minnesota Press, 2001.

Valcheva, Mariya. "Playlistism: A Means of Identity Expression and Self-Representation." A report on scientific research conducted in the Mediatized Stories project at the University of Oslo, 2009.

Vance, James E., Jr. *The Continuing City: Urban Morphology in Western Civilization*. Baltimore: Johns Hopkins University Press, 1990.

———. "Human Mobility and the Shaping of Cities." In *Our Changing Cities*, edited by John Fraser Hart. Baltimore: Johns Hopkins University Press, 1991.

Van Den Eede, Yoni. "Exceeding Our Grasp: McLuhan's All-Metaphorical Outlook." In *Finding McLuhan: The Mind, The Man, The Message*, edited by Jaqueline McLeod Rogers, Tracy Whalen, and Catherine G. Taylor. Regina, Alberta: University of Regina Press, 2015.

Van Der Doelen, Jaap. *Kill Your Masters: Run the Jewels and the World That Made Them*. Athens: University of Georgia Press, 2024.

VanderMeer, Jeff. *Acceptance*. New York: Farrar, Straus & Giroux, 2014.

Van Dijck, José. *The Culture of Connectivity: A Critical History of Social Media*. New York: Oxford University Press, 2013.

———. "Remembering Songs Through Telling Stories: Pop Music as a Resource for Memory." In *Sound Souvenirs: Audio Technologies: Memory and Cultural Practices*, edited by K. Bijsterveld and J. van Dijck. Amsterdam: Amsterdam University Press, 2009.

Van Gennep, Arnold. *The Rites of Passage*. Chicago: University of Chicago Press, 1960.

Viola, Bill, *Reasons for Knocking at an Empty House: Writings 1973–1994*. Cambridge, Mass.: MIT Press, 1995.

Virilio, Paul. *The Lost Dimension*. New York: Semiotext(e), 1991.

Vroon, Piet. "Man-Machine Analogs and Theoretical Mainstreams in Psychology." In *Current Issues in Theoretical Psychology*, edited by W. J. Baker, M. E. Hyland, H. Van Rappard, and A. W. Staats. Amsterdam: Elsevier, 1987.

Wagner, Roy. *Symbols That Stand for Themselves*. Chicago: University of Chicago Press, 1986.

Waite, C. Kaha. *Mediation and the Communication Matrix*. New York: Peter Lang, 2003.

Walker, Rob. *Buying In: What We Buy and Who We Are*. New York: Random House, 2009.

Wallace, David Foster. *Both Flesh and Not: Essays*. New York: Houghton Mifflin Harcourt, 2007.

Wallace-Wells, David. "William Gibson: The Art of Fiction, No. 211." *Paris Review*, Summer 2011. https://www.theparisreview.org/interviews/6089/the-art-of-fiction-no-211-william -gibson.

Walter, E. V. *Placeways: A Theory of the Human Environment*. Chapel Hill: University of North Carolina Press, 1988.

Wampole, Christy. "How to Live Without Irony." *New York Times*, November 17, 2012.

Ward, Graham. Introduction to *The Certeau Reader*, by Michel de Certeau. Oxford, UK: Blackwell, 2000.

Wark, McKenzie. *The Beach Beneath the Street*. New York: Verso, 2011.

———. *Dispositions*. London: Salt, 2002.

———. *50 Years of Recuperation of the Situationist International*. New York: Princeton Architectural Press, 2008.

Watkins, S. Craig. *The Young and the Digital: What the Migration to Social Network Sites, Games, and Anytime, Anywhere Media Means for Our Future*. New York: Beacon, 2009.

Watts, Duncan J., and Steven H. Strogatz. "Collective Dynamics of 'Small World' Networks." *Nature*, June 1998, 393, 440–42.

Waxman, Lori. *Keep Walking Intently: The Ambulatory Art of the Surrealists, the Situationist International, and Fluxus*. Berlin: Sternberg, 2017.

Weheliye, Alexander G. *Phonographies: Grooves in Sonic Afro-Modernity*. Durham: Duke University Press, 2005.

Weick, Karl E., and Karlene H. Roberts. "Collective Mind in Organizations: Heedful Interrelating on Flight Decks." *Administrative Science Quarterly* 38, no. 3 (1993): 357–81.

Weigend, Andreas. "The Social Data Revolution(s)." *Harvard Business Review*, May 20, 2009: http://blogs.hbr.org/now-new-next/2009/05/the-social-data-revolution.html.

Weinberger, David. *Small Pieces Loosely Joined*. New York: Perseus, 2002.

Wells, David. *Dorsality: Thinking Back Through Technology and Politics*. Minneapolis: University of Minnesota Press, 2008.

Wenger, Etienne. *Communities of Practice: Learning, Meaning, and Identity*. Cambridge, UK: Cambridge University Press, 1999.

Wheelwright, Philip. *Metaphor and Reality*. Bloomington: Indiana University Press, 1962.

Wilden, Anthony. *The Rules Are No Game: The Strategy of Communication*. New York: Routledge & Paul, 1987.

———. *System and Structure: Essays in Communication and Exchange*. New York: Tavistock, 1972.

Wiles, Will. *Plume*. London: Fourth Estate, 2019.

Williams, Pharrell. Foreword to *Palm Angels*, by Francesco Ragazzi. New York: Rizzoli, 2014.

Williams, Saul. "Elohim (1972)." In *Black Whole Styles*. CD. New York: Big Dada, 1998.

Wilson, David Gordon. *Bicycling Science*. Cambridge, Mass.: MIT Press, 2004.

Wilson, James Q., and George L. Kelling. "Broken Windows: The Police and Neighborhood Safety." *Atlantic Monthly*, March 1982, 29–38.

Witkin, Herman A., and Donald R. Goodenough. *Cognitive Styles, Essence and Origins: Field Dependence and Field Independence*. Madison, Wisc.: International Universities Press, 1981.

Wolf, Gary. "The Wisdom of Saint Marshall, the Holy Fool." *Wired*, January 1996, https://www.wired.com/1996/01/saint-marshal.

Wolff, Anthony. "Flight: The First 75 Years." *Omni*, December 1978, 47–59.

Wortley, David. *Gadgets to God: Reflections on Our Changing Relationship with Technology*. Leicester, UK: Troubador, 2012.

Yanow, Sophie. *War of Streets and Houses*. Minneapolis: Uncivilized Books, 2014.

Youngblood, Gene. *Expanded Cinema*. Boston: Dutton, 1970.

Zacharias, John. "The Pedestrian Itinerary—Purposes, Environmental Factors and Path Decisions." In *Pedestrian Behavior: Models, Data Collection and Applications*, edited by Harry Timmermans. Bingley, UK: Emerald, 2009.

Zarka, Raphaël. *On a Day with No Waves: A Chronicle of Skateboarding, 1779–2009*. Paris: Edition B-42, 2011.

Zielinski, Siegfried. *Variations on Media Thinking*. Minneapolis: University of Minnesota Press, 2019.

INDEX

Page numbers in *italic* refer to illustrations.

ABOUT THE AUTHOR

Roy Christopher is an aging BMX and skateboarding zine kid. He has written about music, media, and culture for books, blogs, national magazines, and academic journals. He holds a PhD in communication studies from the University of Texas at Austin. His previous books include *Dead Precedents: How Hip-Hop Defines the Future* (Repeater Books, 2019), *Boogie Down Predictions: Hip-Hop, Time, and Afrofuturism* (Strange Attractor Press, 2022), and *The Grand Allusion: A New Understanding of Popular Culture* (Repeater Books, 2025).

www.ingramcontent.com/pod-product-compliance
Lightning Source LLC
Chambersburg PA
CBHW030634221225
37141CB00029B/477